Structure and Function of Plants

Structure and Function of Plants

Jennifer W. MacAdam

A John Wiley & Sons, Ltd., Publication

Edition first published 2009
© 2009 Wiley-Blackwell

Blackwell Publishing was acquired by John Wiley & Sons in February 2007. Blackwell's publishing program has been merged with Wiley's global Scientific, Technical, and Medical business to form Wiley-Blackwell.

Editorial office
2121 State Avenue, Ames, Iowa 50014-8300, USA

For details of our global editorial offices, for customer services, and for information about how to apply for permission to reuse the copyright material in this book, please see our website at www.wiley.com/wiley-blackwell.

Authorization to photocopy items for internal or personal use, or the internal or personal use of specific clients, is granted by Blackwell Publishing, provided that the base fee is paid directly to the Copyright Clearance Center, 222 Rosewood Drive, Danvers, MA 01923. For those organizations that have been granted a photocopy license by CCC, a separate system of payments has been arranged. The fee codes for users of the Transactional Reporting Service are ISBN-13: 978-0-8138-2718-6/2009.

Designations used by companies to distinguish their products are often claimed as trademarks. All brand names and product names used in this book are trade names, service marks, trademarks or registered trademarks of their respective owners. The publisher is not associated with any product or vendor mentioned in this book. This publication is designed to provide accurate and authoritative information in regard to the subject matter covered. It is sold on the understanding that the publisher is not engaged in rendering professional services. If professional advice or other expert assistance is required, the services of a competent professional should be sought.

Library of Congress Cataloging-in-Publication Data

MacAdam, Jennifer W.
  Structure and function of plants / Jennifer W. MacAdam. – 1st ed.
    p. cm.
  Includes bibliographical references and index.
  ISBN 978-0-8138-2718-6 (pbk. : alk. paper) 1. Plant anatomy–Textbooks.
2. Plant physiology–Textbooks. I. Title.
  QK641.M33 2009
  571.2–dc22
                                                            2008035480

A catalogue record for this book is available from the U.S. Library of Congress.

Set in 10/12.5 pt Sabon by Aptara® Inc., New Delhi, India
Printed in Singapore by Fabulous Printers Pte Ltd

2 2009

Dedication

This book is dedicated to scientists in the fields of plant physiology and plant anatomy who continue to uncover how plants work, particularly my colleagues and mentors who communicate their unflagging enthusiasm for these subjects through their own excellence in teaching and research.

Contents

Preface ix
Acknowledgments xi

1 The plant cell 1
2 Plant meristems and tissues 18
3 Plant roots 36
4 Plant stems 55
5 Plant leaves and translocation 71
6 Reproduction in flowering plants 89
7 Plant nutrition 106
8 Plant–water relations 124
9 Macromolecules and enzyme activity 141
10 Photosynthesis 158
11 Respiration 178
12 Environmental regulation of plant development 195
13 Hormonal regulation of plant development 215
14 Secondary plant products 235

Glossary 255
References 273
Index 277

Preface

This book is written for anyone who is curious about plants and wants to better understand the plants we use in our fields and gardens for food, and the plants we love for the beauty they add to our lives. It addresses not only what plants do, but why, to provide insight for our interactions with the plants we cultivate, struggle with, and depend on in the natural environments that are our heritage and our future.

This book is intended for casual readers who want to know how plants actually work, not just how to care for a particular plant in a particular climate, and for students of plant science seeking an understanding of plant structure and function. It is hoped that both groups will be well-served by a simple but accurate overview of the subjects of plant anatomy and plant physiology.

The early chapters of the book are descriptive of roots, stems, leaves, and flowers, and the cells and tissues from which these structures are built. Such organs have much in common from plant to plant, but observing their differences helps us understand plants' adaptations to the often challenging environments in which they developed. The balance of the book describes how plants work: how they extract water and nutrients from the soil, how they use sunlight to create the carbohydrates that are the first link in the food chain, and how they manage to survive and reproduce literally rooted in one spot.

The emphasis of this book is on inclusive explanations, and concepts are illustrated with examples from agricultural and horticultural plants. When used as a text, this material should be supplemented with a hands-on lab where students can observe the anatomical features and physiological processes introduced here. The book is arranged in 14 chapters, a useful length for a one-semester introductory course in plant anatomy and physiology.

Acknowledgments

I thank the many colleagues who generously allowed me to use their drawings and photographs, and the publishers who facilitated my sharing of research figures from the work of numerous other scientists with readers of this book. Adam Black turned sketches of concepts and mechanisms into the illustrations used in this book, and I thank him for his talent and patience.

Structure and Function of Plants

Chapter 1
The plant cell

We appreciate plants for their beauty and usefulness, and on a different level, for the ability of plant species to adapt to an amazing diversity of climates and soils (two of many abiotic influences) as well as their ability to interact with microbes, animals, and other plants (biotic influences). The differences in characteristics such as stem, leaf, and flower structure that result from these and other adaptations were the original basis for classification of plants into different taxonomic groups. However, for all their differences in overall appearance (morphology), plants have the same basic structures at the cellular level, so we begin by looking at the cellular structures of plants. Figure 1.1 is a simplified illustration of a plant cell, and the structures labeled in Figure 1.1 are discussed in more detail in this and other chapters.

Protoplast

The **protoplast** is a collective term that includes the plasma membrane and the cellular objects it contains. It is filled with liquid, the cytosol, that bathes the cellular organelles including the nucleus. The protoplast includes all the "living" parts of the cell, so the cell wall to its outside is not included. The protoplast is composed of 60–75% proteins by dry weight.

Cytoplasm

The **cytoplasm** is the protoplast minus the nucleus. The nucleus directs the work that goes on in the cytoplasm.

Cytosol

The **cytosol** is the liquid portion (matrix) of the cytoplasm, which surrounds organelles and in which a number of proteins, salts (including nutrient ions),

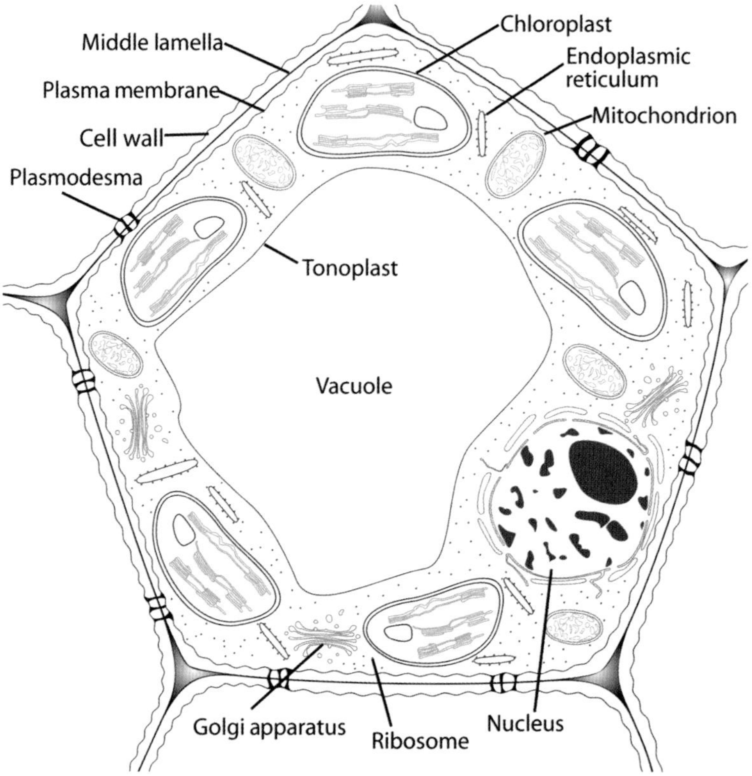

Figure 1.1 Components of a plant cell.

and sugars are dissolved. The cytosol has the thickened consistency of a gel. The cytosol of adjacent cells is continuous, by way of plasmodesmata.

Cell walls

The plant cell protoplast is enclosed by a fibrous wall that grows as the cell expands to its mature size, but which becomes cross-linked and eventually limits the growth of the cell, defining and supporting the cell and collectively providing support for stems and leaves. Some cells, like photosynthetic and storage cells, only have a thin **primary cell wall**, and other cells have both a primary wall and a thick, lignified and therefore rigid **secondary cell wall**, either to retain the cell's shape against the tension of water movement through the plant, as in xylem cells, or to provide concentrated regions of support or protection as in fiber cells or sclerids. The trunk of a tree is made up of concentric layers of water-transporting (xylem) cells with secondary walls that serve both water-carrying and support functions.

Components of the cell wall

Cellulose

The fundamental component of cell walls is **cellulose**, which in turn is made up of long chains of glucose molecules, from thousands to tens of thousands of glucose units per molecule of cellulose. The chemical structure of glucose is illustrated in Figure 1.2, with each of the six carbon atoms (C) numbered. α- and β-Glucose differ in the orientation of the bonds at C-1. Starch and cellulose are both long chains of glucose, but starch is easily digested by monogastrics, like humans, while the bonds between glucose molecules in cellulose are most commonly broken by enzymes produced by microbes inhabiting the guts of ruminants, such as cattle and sheep (and termites). The difference in these chains of glucose is illustrated in Figure 1.3. Bonds in both starch and cellulose are between the 1- and 4-carbons of successive glucose molecules, but while in starch the orientation of each α-glucose molecule in the chain is the same, in cellulose every other β-glucose molecule is flipped on its horizontal axis.

Cellulose is the "fiber" in paper. Cellulose molecules are grouped together into microfibrils consisting of 50–60 cellulose molecules held together by hydrogen bonds, which are relatively loose bonds but effective in large numbers, as in cellulose (Figure 1.4). Cellulose is such a big molecule that it is synthesized at the plasma membrane rather than inside the protoplasm. Microfibrils are extruded into the extracellular matrix, like toothpaste from a

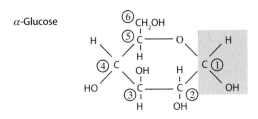

Figure 1.2 The structures of α- and β-glucose, demonstrating the difference in orientation of the –OH at Carbon 1.

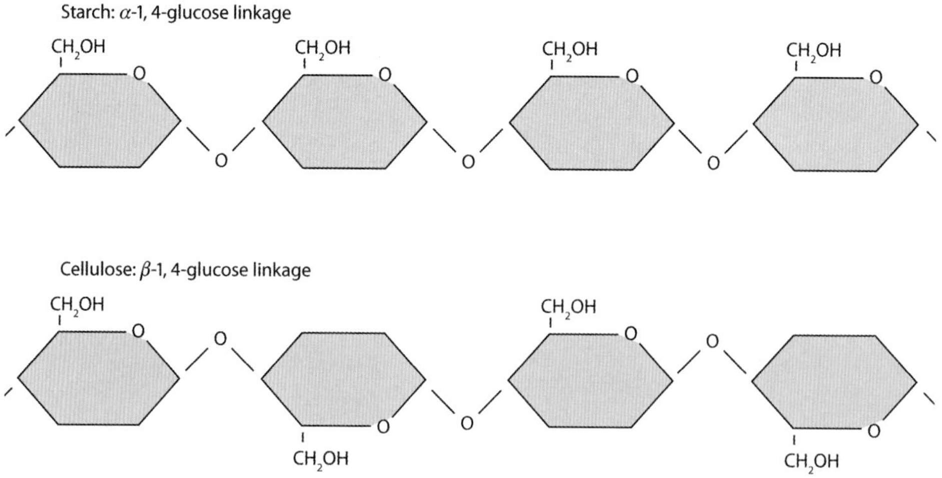

Figure 1.3 The structure of starch and cellulose molecules, demonstrating the difference in bonds between the 1- and 4-carbons of α-glucose molecules in starch and the 1- and 4-carbons of β-glucose molecules in cellulose.

tube (Figure 1.5). Other cell wall components are secreted into the cell wall by way of Golgi vesicles, and assemble around the cellulose microfibrils.

Hemicellulose

Hemicellulose also consists of chains of sugars, but the sugars are much more diverse than in cellulose, which contains only glucose. Hemicelluloses are highly branched because of the bonds that form among the sugars that make them up, and they form a network that coats the much larger cellulose microfibrils. Hemicelluloses adhere to cellulose by way of hydrogen bonds. Hemicellulose molecules coating individual cellulose microfibrils become cross-linked or bound together by covalent bonds, which limits cell wall expansion because the cellulose microfibrils can no longer slide past each other and allow the cell wall to grow. In Figure 1.6, the components of the cell wall are illustrated to show hemicelluloses forming cross-linkages between cellulose microfibrils.

Pectin

The **middle lamella** is the outermost layer of a plant cell and has a high concentration of pectins, which consist of uronic acids, the acidic (and therefore charged) forms of glucose and galactose, and other sugars. The middle lamella is the first boundary formed between what will become adjacent

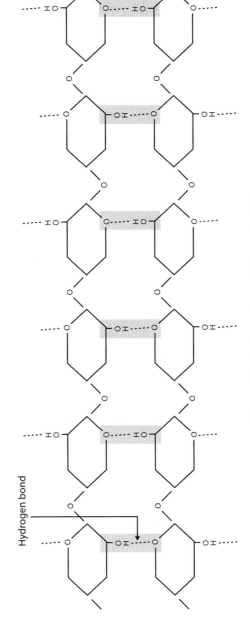

Figure 1.4 Hydrogen bonds between cellulose molecules that result in cellulose microfibrils.

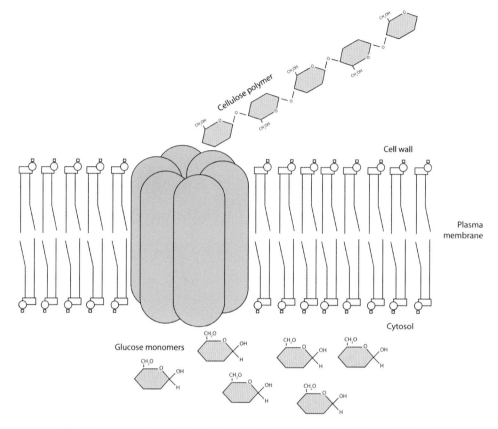

Figure 1.5 Cellulose is formed from glucose by a protein complex, cellulose synthase, at the plasma membrane.

cells during cell division. In cell division, the genetic material of the cell is duplicated and the two groups of chromosomes move to opposite ends of the cell. In Figure 1.7, the middle lamella (yellow) is beginning to form as the boundary between the two new cells. The phragmoplast, a remnant of the organizing structure needed to divide the genetic material, which is shown as groups of white cylinders, is oriented between the daughter nuclei and the developing middle lamella.

After cell division, the primary cell wall forms to the *inside* of the middle lamella, and also has a relatively high content of pectin (up to 35%). The secondary cell wall, when present, then forms to the *inside* of the primary cell wall. The components of both walls are formed in the protoplast and secreted via the Golgi apparatus across the plasma membrane.

Extensin

A structural protein (in contrast to enzymes, which are soluble in the cytoplasm or the matrices of the cellular organelles), extensin, forms a network

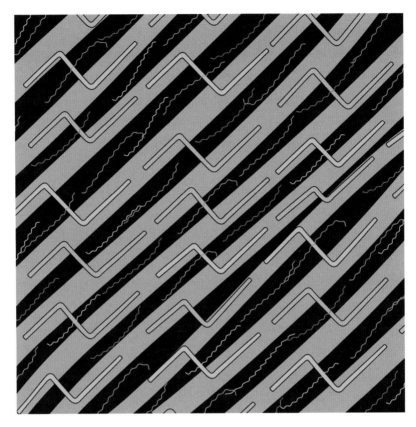

Figure 1.6 Components of the primary cell wall include cellulose microfibrils (gray), hemicelluloses (blue and green), and pectins (orange). In cells such as fibers with secondary cell walls, the space between cellulose and other molecules (black) becomes filled with lignin in both the primary and secondary walls.

within the cell wall that can become cross-linked, like the hemicellulose network. Extensins make up, at most, about 10% of the cell wall, and were first identified in broadleaf plants (dicots), but proteins with similar functions are found in the grasses (monocots).

Secondary cell walls

In structural cells like fibers and in the water-carrying xylem cells, additional cell wall layers are laid down inside the primary cell wall after cell growth stops. These secondary cell walls have a higher cellulose content than primary walls, and may be distinctly layered. In Figure 1.8, which is an electron micrograph of fiber cells, the middle lamella (ML), primary cell wall (CW_1), and distinct layers of the secondary cell wall (S_1, S_2, and S_3) can be seen. The walls of these cells also become lignified, a process in which small **lignin** precursor molecules are secreted into the cell wall and assemble into large,

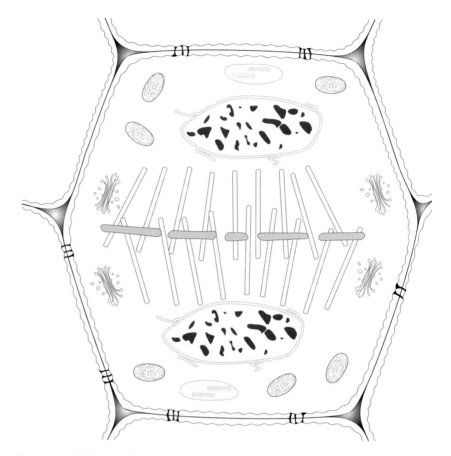

Figure 1.7 Cell division following the duplication of genetic material begins with the formation of a cell plate that develops into the new middle lamella, plasma membranes, and cell walls.

unorganized molecules that displace water (see Chapter 14). The function of lignin is to waterproof xylem vessels and to make cell walls resistant to degradation by invading pathogens. Lignin also greatly increases the rigidity of the cell wall, and is therefore an important component of wood. However, lignin must be extracted for the production of paper, and greatly reduces the digestibility of the fiber cells in plants such as the grasses used as animal feed. Xylem cells, which are the water-carrying cells in roots and shoots, and fiber cells do not contain a protoplasm at maturity and therefore are nonliving cells.

Plasma membrane/cytoplasmic membrane/plasmalemma

The **plasma membrane, cytoplasmic membrane,** and **plasmalemma** are all accepted names for the selectively permeable membrane that encloses the

Figure 1.8 Transmission electron micrograph of fiber cells from the stem of Canada yew (*Taxus canadensis*). Fibers provide structure and protection in leaves, stems, and roots. These cells develop secondary wall layers (S_1, S_2, S_3) inside the primary cell wall (CW_1). The middle lamella (ML) can be seen as a dark line between cells (plate 6.1, p. 98, Ledbetter and Porter 1970, used with kind permission of Springer Science and Business Media).

living contents of the cell and controls the movement of materials into and out of the cell.

Membranes

Plant cell membranes are primarily made of lipids (fats and oils) and proteins. Membranes are usually described as consisting of a **lipid bilayer** because of the way the lipid molecules are arranged, but many proteins are embedded

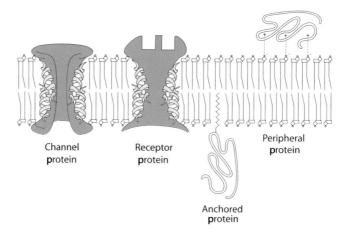

Figure 1.9 The lipid bilayer of the plasma membrane self-assembles from phospholipids, which have hydrophilic, glycerol-containing "heads" oriented outward, and hydrophobic fatty acid "tails" oriented inward. Membrane proteins may extend through the lipid bilayer to act as channels or receptors (green), or they may be bonded to or embedded in the inner or outer surface.

in this bilayer (Figure 1.9). In many cases, these proteins act as gateways for regulation of the contents of the cell.

Membrane lipids

The dominant lipids in the plasma membrane are **phospholipids**, which have a central glycerol molecule with a phosphate molecule attached at one end (the "head") which is water-loving (hydrophilic) and a water-fearing (hydrophobic) "tail" composed of two fatty acids (making these phospholipids diglycerides). The fatty acids are partly unsaturated, making the lipid bilayer fluid, like oils (see Chapter 9).

Phospholipids spontaneously self-assemble into a bilayer in aqueous solutions like a plant cell. They turn their hydrophilic heads outward, some toward the cell wall that encloses the plasma membrane, and some toward the aqueous cell protoplasm, and turn their hydrophobic tails inward to form a double layer. Membranes are fluid—the molecules they contain can easily move past each other in the membrane—but they are also very stable. Membranes can exclude most charged molecules, like nutrient ions, which allows them to control movement of these nutrients into and out of the cell. Water and the gases oxygen and carbon dioxide, however, can cross the lipid bilayer relatively easily.

Other membranes, especially the internal membranes of the chloroplast, contain a large amount of glycolipids, where the head group contains one or two molecules of the sugar galactose instead of a phosphate, and sulfolipids, with a sulfate instead of a phosphate as part of the head group. In these cases, as for phospholipids, the heads are hydrophilic.

Membrane proteins

Proteins make up as much as 50% of the mass of cell membranes. The amino acid composition of proteins determines how the protein is incorporated into the lipid bilayer. If the protein spans the membrane from inside to out, it is an integral protein. If it is bound only to the inside or outside of the bilayer, then it is a peripheral protein. Proteins must have a region that is hydrophobic to be incorporated into the membrane. These membrane proteins can function in the selective transport of solutes across the membrane if they fully span the lipid bilayer (Figure 1.9), or they can act as enzymes like cellulose synthase, or they may form part of an electron transport chain, which are groupings of many different enzymes that are used in photosynthesis and respiration.

Plasmodesmata

The **plasmodesmata** are narrow channels between cells through which dissolved substances but not organelles can pass. Plasmodesmata form during cell division, and allow cell-to-cell communication and transport. One is termed a plasmodesma (Figure 1.10). Plasmodesmata are lined with extensions of the plasma membrane and have an inner structure, the desmotubule, which is continuous with endoplasmic reticulum of the two adjacent cells. Dissolved substances can pass between the plasma membrane and the desmotubule to move between cells. The cytoplasm of adjacent cells connected by plasmodesmata forms a continuous living network among cells called the **symplast**.

In contrast, the **apoplast** is the nonliving space outside the protoplast and includes the cell wall, the intercellular space, and xylem tissue through which water is transported. The larger spaces between cells in leaves and stems is usually filled with air, although cells are coated with a film of water; in roots, these spaces between cells can contain water being taken up by the plant.

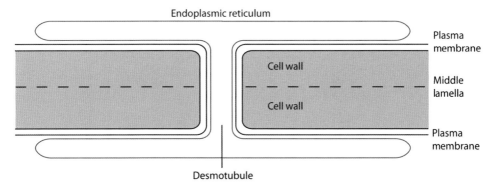

Figure 1.10 Plasmodesmata provide channels between adjacent cells through which dissolved substances can pass.

Cellular organelles

Organelles are membrane-defined compartments inside the cell, each with specific functions. The following are major plant cellular organelles.

Nucleus

The **nucleus** is the location of the genetic material (DNA, deoxyribonucleic acid) contained in almost all cells; the sieve tubes of the phloem are one exception. The nucleus directs the synthesis of the majority of enzyme production and is therefore considered the control center of the cell, since enzymes perform the work (or metabolism) of cells. DNA is organized into chromosomes in plants, and genes are discrete regions of DNA within chromosomes. DNA is the template used to synthesize RNA (ribonucleic acid), which is termed "transcription." RNA is exported from the nucleus to the cytoplasm, where it directs the synthesis of proteins, which is termed "translation." These processes are discussed further in Chapter 9.

Vacuole

Defined by a membrane called the **tonoplast**, the **vacuole** is filled with water, and may comprise 80 or 90% of the volume of a mature plant cell. The vacuole enlarges during growth, and this enlargement occurs by water uptake. The vacuole contains dissolved salts, sugars, organic acids, enzymes, and may contain pigments. Vacuolar transport processes are illustrated in Figure 1.11. The energy of a phosphate bond from ATP (see Chapter 11) is used to pump hydrogen ions (H^+ or protons) into the vacuole; the higher concentration of H^+ in the vacuole reduces the pH of the vacuole compared to the cytosol (for a discussion of pH, see Chapter 7). These H^+ can be exchanged for other positively charged ions such as calcium (Ca^{2+}) and balanced by the

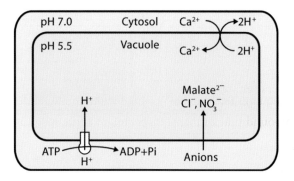

Figure 1.11 The vacuole functions as storage for nutrients, salts, and organic acids, and is active in other transport processes of the cell.

uptake of negatively charged ions such as chloride (Cl^-), nitrate (NO_3^-), or the organic acid malate.

Endoplasmic reticulum

The **endoplasmic reticulum (ER)** is a tubular network that is formed from and continuous with the nuclear envelope, and which fills much of the volume of the cytosol. In Figure 1.1, slices through the ER are indicated as ovals covered with red dots that represent ribosomes. Ribosomes may also occur free in the cytosol. The function of the ER is the synthesis of lipids and proteins that are either used to make cellular membranes or exported from the cell. The space enclosed by the membrane layers is called the **lumen** of the ER. The smooth ER, without ribosomes, is involved in lipid synthesis. The rough ER, with **ribosomes**, is the site of protein synthesis. Proteins that will leave the cell are made on the rough ER and passed into the lumen of the ER after synthesis. In the ER lumen, they are altered by posttranslational modifications such as the addition of sugars to form glycoproteins, which help determine the specific function and location of the protein. The modified proteins move through the lumen to the smooth ER and small, enclosed pieces of the smooth ER bud off to form transport vesicles, with the proteins inside. The transport vesicles move to the Golgi apparatus.

Golgi apparatus (formerly dictyosomes)

A **Golgi apparatus** consists of a stack of separate flattened sacs (cisternae) where products of the ER are processed further. Transport vesicles from the ER deliver their contents by fusing with a membrane of the Golgi apparatus. After processing, products of the Golgi are packaged in secretory vesicles that bud off for transport within the cell, or fuse with the plasma membrane to secrete their contents outside the cell (Figure 1.12). Some of the carbohydrates that make up the cell wall are also formed in the Golgi.

Mitochondria

Mitochondria are the site of respiration (Chapter 11), which releases the chemical energy stored in food (most commonly carbohydrates and fats) and transfers this energy to form ATP (adenosine triphosphate), a chemical compound that can be transported to other locations in the cell. The outer membrane of a mitochondrion is readily permeable, and the inner membrane, which has many invaginations called cristae, is very high in protein (70%), and is much more selective. Inside the inner membrane is a matrix that has a high concentration (40–50%) of dissolved protein (enzymes) where key steps in respiration occur (Figure 1.13).

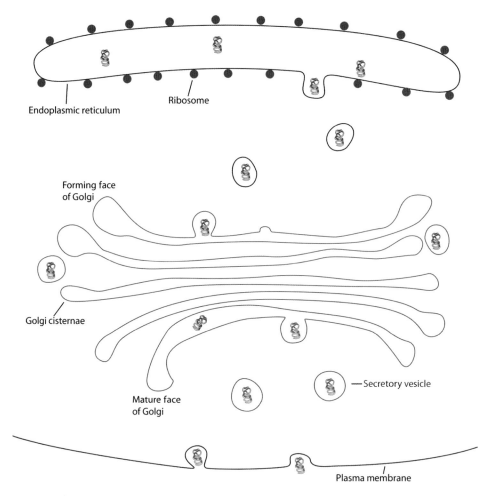

Figure 1.12 The Golgi apparatus functions in the processing and export from the cell of glycoproteins (proteins with sugars attached) synthesized in the endoplasmic reticulum (ER), and in the synthesis and export of some cell wall carbohydrates (hemicelluloses and pectins).

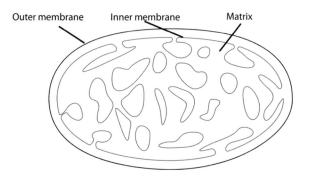

Figure 1.13 Mitochondria are the cellular organelles in which most steps of respiration occur.

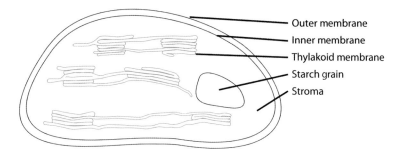

Figure 1.14 Chloroplasts are the cellular organelles in which photosynthesis occurs.

Chloroplasts and other plastids

Chloroplasts are football-shaped organelles found primarily in mesophyll cells of leaves and stems and guard cells of the epidermis. These organelles are the site of photosynthesis, the metabolic process that synthesizes sugars in plants. Chloroplasts consist of three membranes: a readily permeable outer membrane, a selective inner membrane, and the thylakoid membrane system. The matrix that fills the space around the outer surface of the thylakoid membranes is called the stroma, and is protein-rich (Figure 1.14). The reactions of photosynthesis that produce sugar occur in the stroma. Chlorophyll is a pigment that is embedded in the thylakoid membranes, which enclose a space called (as with the ER) the lumen. The lumen is where water is oxidized for the generation of the O_2 that is a by-product of photosynthesis, and where protons (hydrogen ions; H^+) accumulate, subsequently driving ATP synthesis. Some thylakoid membranes occur in stacks called grana. Other plastids are the storage sites for starch (amyloplasts; common in roots; Figure 2.4) and for pigments (chromoplasts), such as those in carrot roots, tomato fruits, and the petals of yellow, orange, or red flowers.

Endosymbiosis

Both chloroplasts and mitochondria are thought to have originated as bacteria (prokaryotic cells that have no nucleus; Figure 1.15a) that invaded or were consumed by eukaryotic cells, which do have nuclei (Figure 1.15b), forming a symbiotic (mutually beneficial) relationship (Figure 1.15c). The invasion by the bacteria that became chloroplasts (Figure 1.15d) followed the invasion by the bacteria that became mitochondria (Figure 1.15e), creating plant cells (Figure 1.15f). Our understanding of endosymbiosis was developed over many years by Dr Lynn Margulis. A unique feature of mitochondria and chloroplasts among plant cell organelles is the presence of inner and outer membranes in both organelles. The outer membrane is thought to have formed from an invagination and eventual budding off of the plasma membrane as the cell tried to contain an invasion by the bacterium. In addition, both organelles contain plasmid (bacterial-type) DNA, and still use it to carry out their own protein synthesis. These invaders benefited from their new surroundings, and were thought to have been tolerated because their presence confers clear advantages to the cells that contain them. Therefore, the relationship is mutually beneficial or symbiotic.

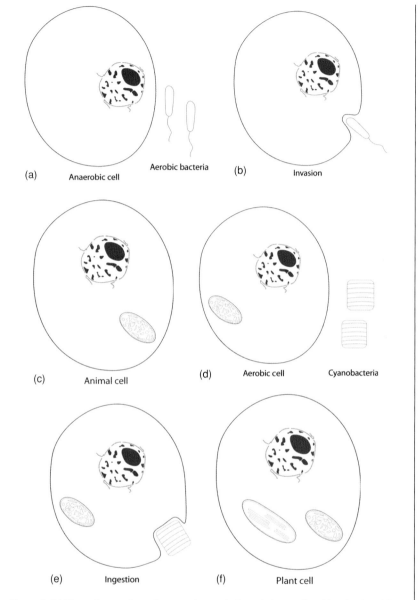

Figure 1.15 These figures show the stepwise evolution of plant cells with mitochondria and chloroplasts via endosymbiosis. (a) Cells exist that have a nucleus but which cannot carry out aerobic respiration or photosynthesis. (b) The cell with a nucleus is invaded by a cell that is capable of aerobic respiration. (c) Rather than digesting the invading bacterium, the cell and the bacterium provide benefits to each other. (d) Other bacteria are capable of photosynthesis. (e) The eukaryotic cell (with a nucleus) may have consumed the bacterium capable of photosynthesis. (f) The further benefits of a resident "organelle" carrying out photosynthesis causes a further symbiosis. This was the basis for the organisms that developed into plants. The bacterial precursor of chloroplasts allowed the new cells to produce sugars, and the bacterial precursor of mitochondria allowed the new cells to produce energy from these sugars much more efficiently.

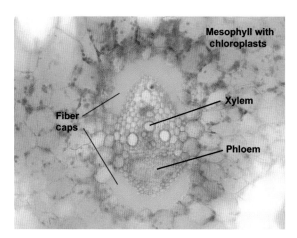

Figure 1.16 A vascular bundle from a pineapple (*Ananas comosus*) leaf surrounded by cells for photosynthesis (mesophyll). The vascular bundle contains cells for the transport of water (xylem) sugars and amino acids (phloem). The two fiber caps support the leaf and protect the transport tissues.

Variation in cellular structure

Plant cells have many basic structures in common, but as can be seen in the vascular bundle in Figure 1.16, variation in these structures can result in enormous differences in cellular appearance and function. Most remarkably, these very different cells are connected to one another through plasmodesmata; they divide and expand together as plants develop and work together through the life of the plant.

Chapter 2

Plant meristems and tissues

Plants are usually described in terms of their stem height, their leaf shape, and their flower shape and color. These plant organs are in turn composed of tissues, and grow into their final form through cell division, expansion, and differentiation. In this chapter, the regions of cell division (meristems) that develop into plant organs are described. Tissues are also described in the general terms used by botanists, which apply to similar cells wherever they occur in the plant. In the chapters that follow, we explore how these different types of tissues—specialized for protection, support, or transport—are organized into roots, stems, leaves, and flowers.

Meristems

In contrast to animals that attain a mature size and stop growing, plant roots and stems continue to grow throughout the life of the plant. Perennial plants produce new branches, leaves, and flowers each year. **Meristems** are regions of cell division that produce new cells for primary and secondary growth in all tissues of the plant. **Primary growth** results in an increase in the length of shoots and roots, including branches, while **secondary growth** results in an increase in the girth of woody stems and roots.

Plant meristems

Apical meristems

Apical meristems are located at the tips (apex) of stems and roots, and leave cylindrical roots or stems behind them as they produce new cells. The apical meristem of a grass plant is located at the top of the stem, but if the stem is not elongated, the apical meristem is enclosed in leaves (Figure 2.1). In

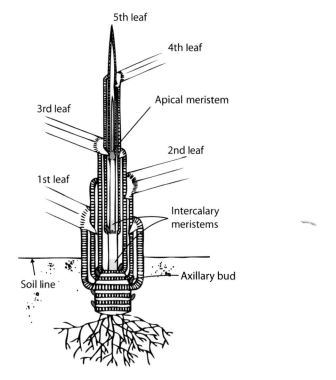

Figure 2.1 Diagram of a grass plant with four mature leaves comprising horizontal blades and vertical sheaths, and a fifth elongating leaf emerging through the whorl of older leaves. Leaves are attached to an elongating stem. The apical meristem is green, the intercalary meristems are yellow, and the axillary buds are blue (Dodds 1980).

meristems, cells divide, increase in size, and divide again. Apical meristems are usually protected by the plant; they are covered by a root cap in the soil or enclosed by leaves, bracts, or scales on the shoot (Figure 2.2). There are apical meristems at the tips of all living roots; just a few are indicated by red arrows in Figure 2.3.

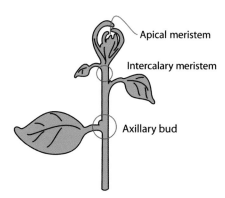

Figure 2.2 Diagram of the shoot of a broadleaf plant showing an axillary bud at the junction of a leaf petiole and the stem, an intercalary meristem located at the base of a stem internode, and the apical meristem enclosed in young leaves.

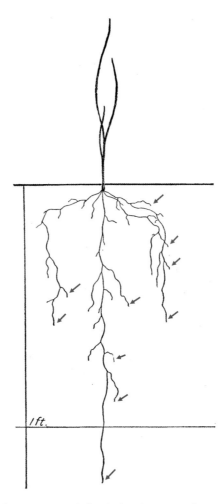

Figure 2.3 Diagram of a young onion (*Allium*) plant (Weaver and Bruner 1927) with the location of some of the many root apical meristems noted with red arrows.

In both roots and shoots, there are slowly dividing groups of **initials**, the cells that constitute a meristem for the apical meristems. These groups of initials are called the **quiescent center** in roots and the **central mother cells** in shoots. Infrequent divisions by initial cells produce the cells that divide more actively to produce the primary plant body. Mutations are more likely to occur the more often a cell divides, so if a frequently dividing cell mutates, it can be replaced from the initials "storehouse" where genetic information is still correct.

Axillary buds

Axillary buds are located in leaf axils, the junctions of the leaf, and the stem (Figures 2.1 and 2.2). The activity of these meristems is similar to that of apical meristems, but they produce branches or flowers. The pattern of branching of a stem is established by the axillary buds produced by apical meristems, and regulated by the hormonal activity of the apical meristem.

Intercalary meristems

Intercalary meristems are located between mature tissues, most commonly in stems (Figure 2.2), and leaves of grasses. These meristems are located at the base of leaves and stem internodes; the activity of intercalary meristems results in the regrowth of grasses following mowing or grazing. In Figure 2.1, the intercalary meristems of the grass stem are colored yellow.

Lateral meristems

Lateral meristems form a cylinder around the stems and roots of woody plants and produce secondary growth. One of these is the **vascular cambium**, which produces the secondary xylem and phloem; another is the cork cambium, which produces the cork layer that replaces the epidermis of woody perennial shoots and roots.

Tissues formed during primary growth

1. **Ground tissue** is the group of simple tissues that comprise the bulk of the plant.
2. Epidermal tissue is the "skin" or outer covering of shoots and roots.
3. Vascular tissues are the "plumbing" system, carrying water, amino acids, and sugars throughout the plant.

Ground tissues

Parenchyma

Parenchyma cells have a relatively thin primary wall and form the bulk of many plant parts, like the **cortex** and **pith** of the stems of herbaceous plants, and the cortex of roots. In the taproots of starch-storing plants, starch is stored in cortical cells in membrane-bound organelles called **amyloplasts**. In Figure 2.4, the walls of starch-storing cells are transparent, and amyloplasts can be seen filling the cells. The winter-hardy legume cicer milkvetch (*Astragalus cicer*) stores starch under ground in rhizomes, and amyloplasts can be seen in the iodine-stained rhizome section in Figure 2.5. The **mesophyll** (photosynthetic) tissue of leaves is also parenchyma tissue. Mesophyll cells are an example of ground tissue that has an important function in plant metabolism.

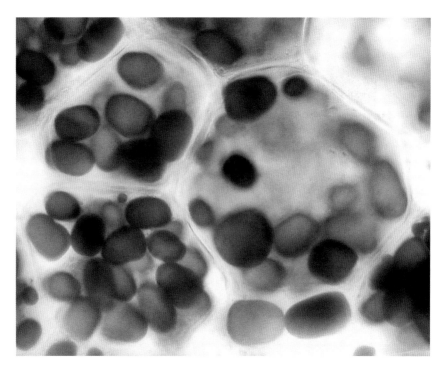

Figure 2.4 Cells from a Dutch iris (*Iris hollandica*) bulb scale filled with amyloplasts and stained for starch with iodine.

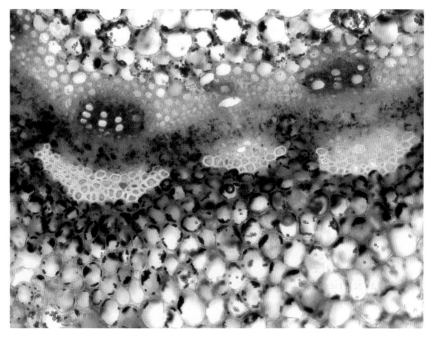

Figure 2.5 Cross section of a cicer milkvetch (*Astragalus cicer*) rhizome stained with iodine. Many of the cells near the vascular bundles contain dark-stained amyloplasts.

Figure 2.6 Cross section of a begonia stem. Collenchyma cells, located just inside the epidermis, have primary walls thickened at triangular cell junctions. Walls are stained dark red in this photograph.

Collenchyma

Collenchyma cells have primary cell walls that are thickened, often at cell junctions, appearing as darkly stained triangles in Figure 2.6. These cells resemble fibers because they are very elongated (up to 2 mm in length) and they provide support, but their walls do not become lignified. Collenchyma cells are found in growing tissue of the shoot, and they are alive at maturity. The stalks of celery (*Apium*) and rhubarb (*Rheum rhabarbarum*) are actually petioles (the stalk of a leaf) that contain strands of collenchyma tissue (the strings in celery) for support. Because these cell walls are not lignified, they are not rigid, and depend therefore on turgidity (water under pressure) to assist in their support function. This lack of lignification is why celery wilts as it dries while a woody tissue does not.

Sclerenchyma

Sclerenchyma cells also provide support, but have thick secondary cell walls inside their primary cell walls that become lignified at maturity, adding

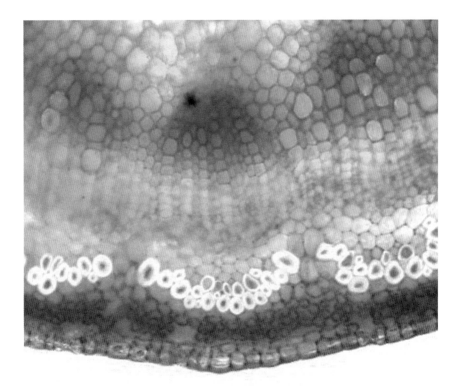

Figure 2.7 Cross section of a young flax (*Linum usitatissimum*) stem. Fiber cells appear as white rings and are grouped in semicircles to the outside of the phloem.

considerable rigidity. Sclerenchyma cells are often dead at maturity. There are two types of sclerenchyma cells: fibers and sclerids. Fibers are long, slender sclerenchyma cells that can sometimes be used to produce cloth like linen, or rope like hemp. Cotton (*Gossypium hirsutum*) fibers, however, are trichomes or extensions of epidermal cells. Fibers often occur alongside vascular tissue in plants, providing protection and support. In Figure 2.7, the walls of groups of fibers of a young flax (*Linum usitatissimum*) stem appear as white rings. The cellulose in **fiber cells** of grasses and other forages can be utilized by rumen microbes in the digestive system of grazing animals, but the greater the lignification of these cell walls, the less digestible the feed is. **Sclerids** are short and of variable shapes; sclerids can be encountered as the grit in pear (*Pyrus communis*) fruit and they form the shell of nuts such as walnut (*Juglans*). Figure 2.8 shows a still-living pear sclerid with a primary (CW_1) and secondary (CW_2) wall. This cell still retains its nucleus (N), mitochondria (M), and plastids (P). There are also islands of the cytoplasm seen in the CW_2 with endoplasmic reticulum (ER). The inset shows the middle lamella and primary wall with plasmodesmata (Pd) coinciding with an area of cytoplasm.

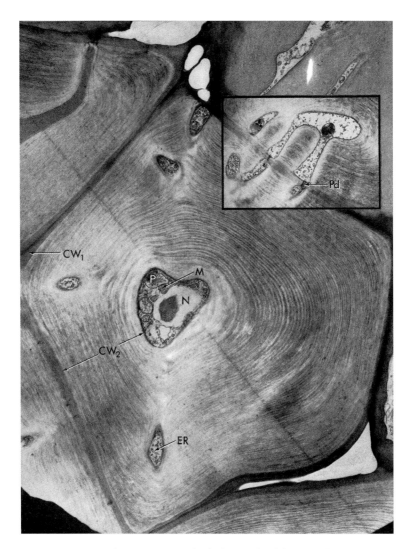

Figure 2.8 Transmission electron micrograph of a living sclerid from pear (*Pyrus communis*) fruit. A primary cell wall (CW_1) encloses a still-forming secondary cell wall (CW_2). The protoplast contains a nucleus (N), mitochondria (M), plastids (P), and endoplasmic reticulum (ER). The inset shows the middle lamella and walls of two sclerids crossed by plasmodesmata (Pd) connecting the cytoplasm of the two cells (Plate 6.2, p. 102, Ledbetter and Porter 1970, used with kind permission of Springer Science and Business Media).

Epidermal tissue

The **epidermis** is a continuous protective layer of cells on the outer surface of both roots and shoots. In the shoot, epidermal cells are coated with a layer of waxy cutin (the **cuticle**) over their outer walls to limit water loss from the surface. In Figure 2.9, the cuticle of yucca appears as a gray layer continuing from the epidermis across guard cells of the stoma. The waxy

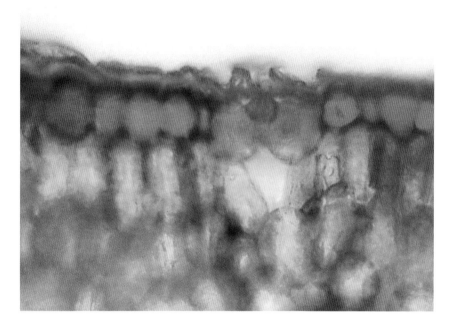

Figure 2.9 Epidermis (red) and stoma of pineapple (*Ananas comosus*) with a thick cuticle (gray) covering the outer surface of the epidermis to prevent evaporation.

cuticle causes water to bead on the surface of leaves, necessitating the use of a surfactant when applying compounds to leaf surfaces. The epidermis is usually just one cell thick except in shoots of succulents such as jade (*Crassula ovata*) or rubber plant (*Ficus elastica*), where the extra epidermal layers are used for water storage. Epidermal cells typically do not have chlorophyll and are transparent, but their vacuoles may contain other pigments such as anthocyanins which absorb ultraviolet light and therefore act as a sunscreen.

Pores in the epidermis are termed **stomata** (one is a stoma) and are composed of two **guard cells** that are specialized epidermal cells that control the movement of CO_2 and, in drought conditions, of water into or out of the leaf. The guard cells are flanked by two or four **subsidiary cells** that assist in stomatal function, and the guard cells, inner air space, and subsidiary cells constitute the stomatal complex (Figure 2.9). The leaves of most plants have several thousand stomata per square centimeter. In dicots, stomata tend to be on the lower sides of leaves, while in grasses, they are more evenly distributed on both the upper and lower leaf surfaces. A replica of the lower epidermis of English ivy (*Hedera helix*; Figure 2.10) shows that the frequency of stomata (arrows) is high relative to the frequency of epidermal cells in dicots.

The swelling of guard cells that leads to stomatal opening is caused by the uptake of potassium (K^+) and other charged substances into guard cells. The guard cells of dicots are "kidney"-shaped, and the stoma opens as the guard cells swell because of the pattern of reinforcement of the cell walls.

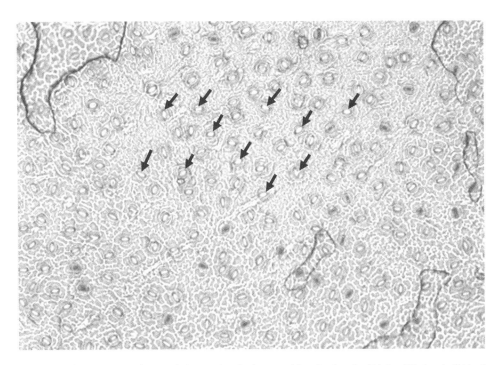

Figure 2.10 Replica made by coating the lower epidermis of an English ivy (*Hedera helix*) leaf with clear nail polish and lifting it from the leaf with cellophane tape. Some of the many stomata are indicated by red arrows.

Monocot guard cells are "dumbbell"-shaped when they are swollen, and open because the ends of the cells enlarge and force the centers of the cells apart. The transparent lower epidermis peeled from a leaf of the grass tall fescue (*Lolium arundinaceum*; Figure 2.11) shows rows of stomata (arrows) that occur on either side of each vein. Also notice the barbs along the margin of the leaf, which act as a defense against grazing. Ledges on some guard cells reduce evaporation by helping to maintain still air immediately above the stomatal opening, and also keep liquid water from entering stomata.

Leaf rolling in grasses is caused by specialized epidermal cells called **bulliform cells** which are located in groups between the ridges formed by veins. In Figure 2.12, a section of an unstained tall fescue leaf, the groupings of transparent bulliform cells (under the red arrow) can be seen between mesophyll cells filled with green-pigmented chloroplasts. Drought stress causes these large, water-filled cells to lose water and shrink, which causes the leaf to roll, minimizing the exposed surface area and reducing further water losses.

Trichomes are specialized epidermal cells in the form of hairs. Cotton fibers are trichomes of cottonseed epidermal cells, which can grow up to 6 cm (2.5 in.) in length. Root hairs such as the maize (*Zea mays*) root hairs in Figure 2.13 are also trichomes. In this case, root hairs increase the surface area for uptake of water in young roots. Trichomes can deter insects from attacking leaves by providing a physical barrier or in some cases by impaling insects

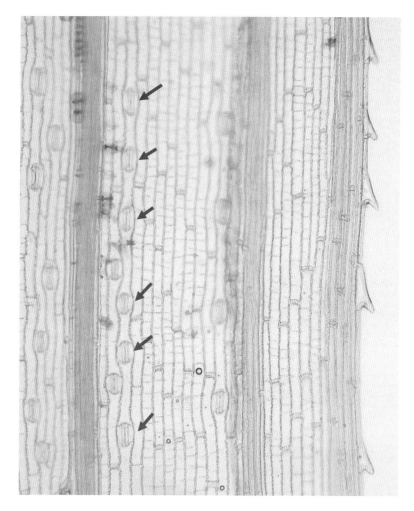

Figure 2.11 Lower epidermis peeled from a leaf of tall fescue (*Lolium arundinaceum*). The stomata in one file of epidermal cells are indicated with red arrows.

that crawl across the leaf. Other trichomes contain irritating chemicals or sticky substances that entrap or scare off attackers (e. g., stinging nettle) and still others send distress signals to attract predators of the plant pest.

Vascular tissues

The vascular tissues allow the organs that carry out photosynthesis above ground (the leaves) to be spatially separated from the organs that collect water and nutrients from the soil below ground (the roots). Vascular tissue may be derived from the apical meristems of the plant, or from the vascular cambium, which is especially important in the growth of wood (i.e., tree trunks).

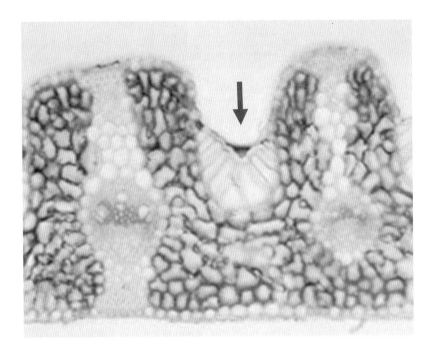

Figure 2.12 A cross section of a fresh, unstained tall fescue (*Lolium arundinaceum*) leaf. Mesophyll cells surrounding each vein are filled with green-pigmented chloroplasts. Large bulliform cells occur in the upper epidermis between two veins.

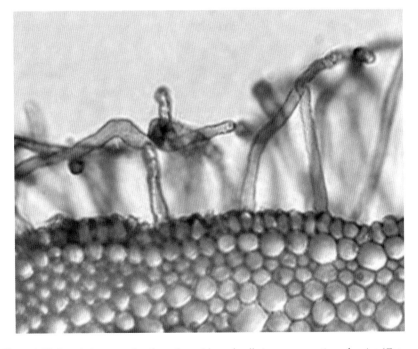

Figure 2.13 Root hairs extending from the epidermal cells in a cross section of maize (*Zea mays*) root. Root hairs are one example of trichomes.

The vascular cambium lays down **xylem** cells to the inside and **phloem** cells to the outside to increase the diameter of woody stems or roots.

Xylem tissue

Water and dissolved mineral nutrients (salts like potassium) are conducted through **tracheids** and **vessel elements** that have thickened, lignified cell walls that waterproof the cell, keep the cell from collapsing under tension, and also provide support. When mature, these water-conducting cells no longer have a protoplast and the water inside is in a column that is continuous from the roots to the leaves. Tracheids, found in both gymnosperms (which have naked seeds) and angiosperms (which produce flowers and fruits), may be more than a centimeter (0.4 in.) in length, and are connected at both their tapered, overlapping end walls and their sides to other water-conducting cells (Figure 2.14a). Water moves from one tracheid into the next through pit pairs along the length of the cells. Pit pairs are interruptions in the cell wall that match one another in adjacent cells. Bordered pits in conifers are capable of sealing the opening between two tracheids with a torus, a thickening of the middle lamella (pit membrane) shared by the two cells. Normally, water rises through the tree by moving through the porous pit membrane (Figure 2.15a) Under drought, water is being lost from plant shoots but not replaced from the dry soil. As tension increases from the pull of transpiration, the water column in the xylem may break, resulting in a bubble in a tracheid. Plants protect themselves from propagation of these air bubbles throughout the xylem by sealing off the tracheid filled with air by closing off the bordered pits (Figure 2.15b). In such a case, low tension on the air side and high tension on the water side pulls the seal (torus) against the water side of the cell wall.

In angiosperms like maple (*Acer*), the leaf surface area is much greater than in gymnosperms (conifers like pines). More photosynthesis can occur in these thin, wide leaves than in the needles of conifers, but more water is transpired from the leaf surface in the process. Water movement is through vessel elements, and is facilitated through openings in their adjacent end walls called **perforation plates** (Figure 2.14b), or their end walls may be completely dissolved away (Figure 2.14c). The wood of angiosperms usually includes both tracheids and vessel elements plus fiber cells and water-storing parenchyma, while gymnosperms have only tracheids. Vessel elements are usually shorter and wider than tracheids, but in some cases may be as long as 30 cm (1 ft) in woody species. Xylem water-carrying cells are continuous and interconnected from root tips to leaves.

Phloem

Phloem tissue carries sugars and amino acids synthesized in leaves to other parts of the plant where growth and other metabolic processes require energy

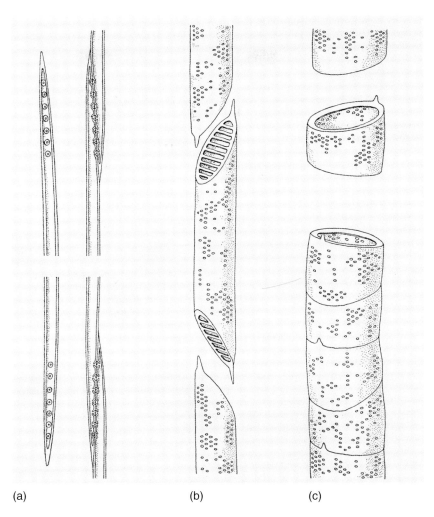

Figure 2.14 Water-conducting cells of the xylem: (a) tracheids with bordered pits from pine (*Pinus*), a primitive conifer; (b) vessel elements from birch (*Betula*), with partially dissolved end walls; (c) segments of large xylem vessels from oak (*Quercus*), without end walls. (Illustration by Irving Geis in Zimmermann 1963, used with permission.)

and building blocks. These molecules dissolved in water are carried through **sieve elements** that are joined end to end at **sieve areas** perforated by sieve pores (or sieve plates). The sieve elements of nonflowering plants are relatively primitive and are called sieve cells. Most angiosperms (flowering plants) have sieve tube members (Figure 2.16). These sieve elements have no nucleus at maturity, but do have cytoplasm, and remain thin-walled and very much alive in contrast to xylem cells. The cytoplasm of sieve tube members contains no vacuole, dictyosomes, or ribosomes, but it is well supplied with mitochondria and has modified endoplasmic reticulum. The cytoplasm

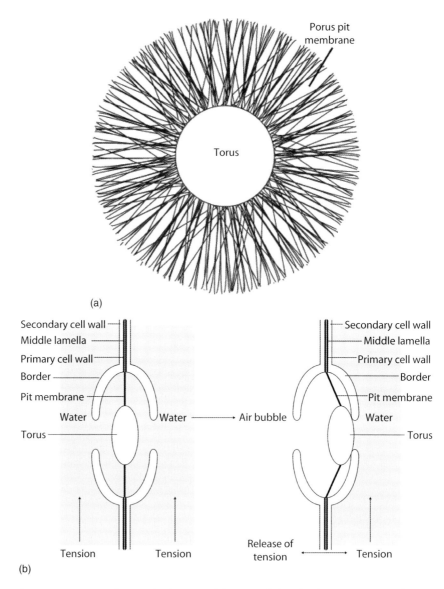

Figure 2.15 Diagram showing the function of the torus of a bordered pit in preventing the propagation of an air bubble into adjacent cells. Water can move from cell to cell through openings in the pit membrane around the torus (a). When an air bubble forms in one cell, the difference in tension pulls the pit membrane to one side, plugging the channel with the torus (b).

of two adjacent sieve elements is connected through the sieve areas. Like the xylem, phloem tissue also includes parenchyma and fiber cells. Some of the parenchyma cells are **companion cells**, derived from the same mother cell as the sieve tube member, and are thought to provide metabolic support through numerous cytoplasmic connections via plasmodesmata. Phloem parenchyma

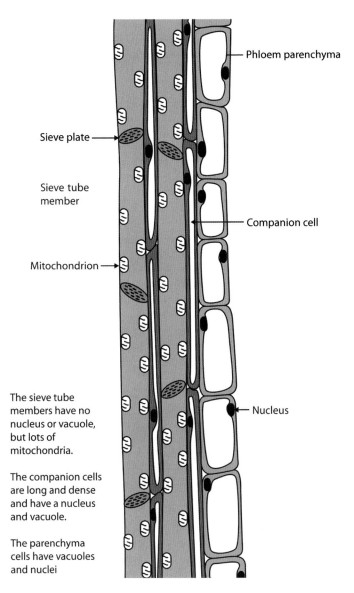

Figure 2.16 The phloem of angiosperms consists of sieve tube members that transport sugars and amino acids but do not contain a nucleus, companion cells that do contain a nucleus to support the function of sieve tube members, and parenchyma cells that function in the transport of sugars from photosynthetic cells into the phloem.

cells are thought to function in the collection and transfer of photosynthate into the phloem.

Like the secondary xylem, the secondary phloem is produced by the vascular cambium. In a tree trunk, xylem forms wood to the inside, and phloem to the outside. All vascular tissues—groups of cells that include both xylem and phloem—typically have the same orientation, with phloem to the outside or down and xylem on the inside or up.

Accidental bonsai

The little juniper at the center of this photo (Figure 2.17) is perfectly formed, but its trunk is only about 6 cm (2.5 in.) tall. What inhibited cell division in the meristems of this tree to change its normal growth habit so drastically? Genetically, it is probably very similar to its larger relatives growing nearby that were the source of the seed from which this little tree developed. One piece of evidence for its normal genetic makeup is that its scale leaves are no smaller than those on other trees.

The art of bonsai originated in China in the seventh century AD, and the original name for it meant "tray scenery." The creation of a bonsai tree is done by pruning the roots of a young perennial woody plant, such as a juniper, pruning the shoot to resemble the branching pattern of a mature tree, and establishing the tree in a shallow tray. Only about an inch of soil is used, so the cultivation of an intentional bonsai tree is done in an environment similar to the one created by nature at Hovenweep National Monument, where the tree in this photograph is growing.

Bonsai trees require frequent watering, good drainage, a steady supply of nutrients, and a normal level of light—all ingredients that are present for this accident bonsai tree: the decaying plant material and seep above the plant will supply nutrients and water, the little ledge on which the tree is growing has good drainage, and the tree

Figure 2.17 A juniper (*Juniperus*) tree whose growth is stunted by the limited volume of soil into which it germinated.

is growing in the open, so it is receiving normal light. The shallow accumulation of soil into which this tree's seed germinated has limited the growth of its roots, thereby naturally pruning the root system. The proliferation of roots into the accumulated soil will help keep the soil in place, and as old roots die and are replaced by new ones, decaying organic matter will add valuable, soil-building humus to the rooting environment, improving the water- and nutrient-holding capacity of the soil.

Accidental bonsai trees like this one are common in the dry southwestern United States, where they provide a testimony to the tenacity of plants and the ingenuity of nature.

Chapter 3

Plant roots

Plant roots have many functions critical to the plant and must survive in an environment that is alternately wet and dry, entirely dependent on the plant shoot for sustenance, seeking water, nutrients, and oxygen in competition with animals, microbes, and the roots of other plants. Compared to leaves, which are thin and flat to maximize interception of sunlight, roots are cylindrical and often very thin to maximize surface area for water absorption. Plant roots respond to gravity, they host fungi and bacteria in exchange for hard-to-find nutrients, and they inform the plant about conditions in the soil via hormones and other signals. In this chapter, the structure and development of the root are described, and in later chapters, nutrient uptake (Chapter 7) and hormonal interactions with the shoot (Chapter 13) are explored.

The functions of roots

1. Roots anchor plants in place in the soil and provide support to keep plants upright.
2. Roots absorb water and dissolved mineral nutrients from the soil, and transport them to the shoot.
3. Taproots are sometimes used for the storage of food reserves (e.g., alfalfa and carrots). In biennial and perennial species, root reserves are the source of carbohydrates and proteins for new growth in the spring.

Types of root systems

Most monocots, such as grasses, have fibrous root systems, an extensive group of relatively thin roots of similar size. Fibrous root systems are often

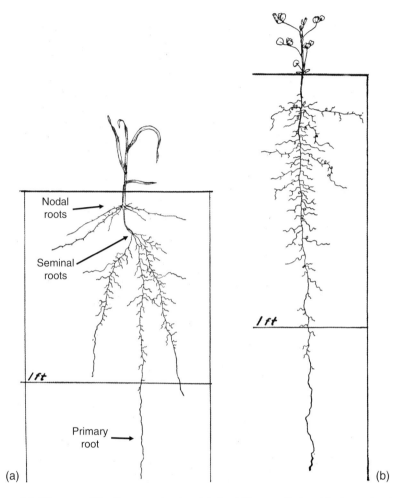

Figure 3.1 Diagrams of the fibrous root system (a) of oat (*Avena sativa*) and the taproot system (b) of alfalfa (*Medicago sativa*). (Drawings from Weaver 1926.)

relatively shallow, and absorb water and nutrients from the upper foot or two of the soil. Because **fibrous roots** are so numerous and concentrated near the soil surface, they are excellent for the prevention of soil erosion. In grasses, the root that develops from the radicle or seed root is the primary root (Figure 3.1a), and seminal roots develop from the base of the first node at the seed. The site of permanent root development in grasses is the base of the unelongated lower stem internodes. Roots that develop from organs other than the root are termed **adventitious roots** and are located near or below the soil surface.

The subunit of a grass tiller is termed a phytomer, and is composed of a leaf blade, leaf sheath, the node to which the leaf is attached, the internode below this node, and roots located above the next node of the stem. Therefore, nodal

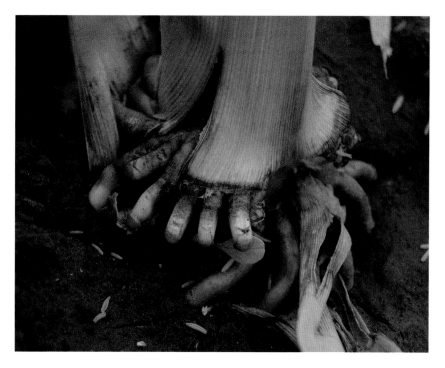

Figure 3.2 The brace roots of maize (*Zea mays*) are adventitious roots that are initiated above the soil surface. When these roots enter the ground, they form lateral branches like other roots.

roots can supply water and nutrients directly to one leaf of a grass tiller. The brace roots of maize are nodal roots that are initiated above ground and then grow into the soil (Figure 3.2).

Most conifers and dicots have **taproot** systems that develop from the radicle. This consists of the taproot itself, and smaller branch and fibrous roots (Figure 3.1b). The development of branch roots is governed by hormones from the taproot in the same way that the development of branches in the shoot is controlled by hormones from the main stem of a plant. If the taproot is damaged, a branch root will enlarge and take over the function of the taproot.

Taprooted plants can explore the soil to great depth to supply the plant with water. Locoweed (*Oxytropis lambertii*) is a taprooted legume native to the short-grass prairie (Figure 3.3). Grasses deplete the surface soil moisture, so locoweed germinates into depressions or rodent burrows where water penetrates into the subsoil (Weaver 1926).

Contractile roots develop on some legume species such as alfalfa (*Medicago sativa*; Figure 3.4), and some bulbs such as gladiolus. These roots contract to pull the shoot underground by expanding radially, or to the sides, while contracting in length. Burying the crown of the shoot underground provides protection from fluctuations in air temperature.

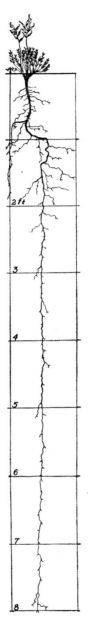

Figure 3.3 A diagram of purple locoweed (*Oxytropis lambertii*) showing the depth of the taproot in feet. (Drawing from Weaver 1926.)

The organization of root growth zones

Root growth occurs at the tip of roots, and the most active region of water and mineral nutrient absorption is also near the tip, in elongated but not fully mature regions of the root.

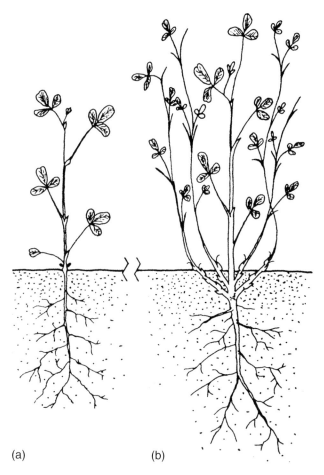

Figure 3.4 Diagram of an alfalfa plant before and after contractile growth. At the seedling stage (a) there are axillary buds at the cotyledonary nodes just above the soil surface, and a unifoliolate and four trifoliolate leaves are present. Following contractile growth (b), branches have developed from the axillary buds at the cotyledonary, unifoliolate, and first trifoliolate leaf nodes, which were pulled underground to form the crown. The depth of crown development in perennial legumes has been positively correlated with winterhardiness. (Drawing from Nelson and Moser 1995, used with permission.)

The root cap

The **root cap** is like a thimble that covers and protects the growing root tip (Figures 3.5 and 3.6a). It has been shown that the root cap also senses light and the pressure exerted by soil particles. While the root cap does not regulate the rate of root elongation, it is the source of the perception of gravity by the root (Juniper et al. 1966). The root cap has its own meristem (region of cell division). Root cap cells are pushed away from the meristem as they enlarge by continuing cell division at the meristem, and in turn, replace older cells that are worn away at the outside edges of the root cap. A root cap can shed thousands of cells each day. The outermost root cap cells as well as epidermal

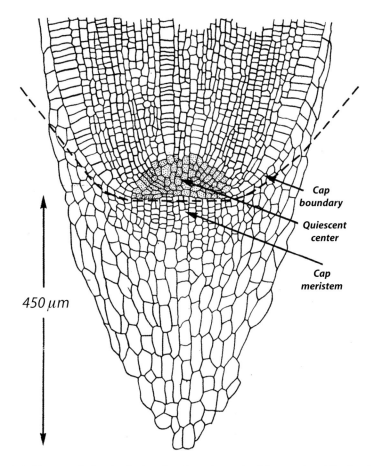

Figure 3.5 Diagram of a longitudinal section through a maize (*Zea mays*) root tip showing the root cap, its meristem, and the quiescent center that supplies initials to both the root cap and root meristems. (Drawing from Juniper et al. 1966, used with permission.)

cells of the root secrete a slimy substance called mucigel, which contains sugars and other compounds produced by the Golgi apparatus. Mucigel lubricates the pathway through the soil for root growth, protects the root from desiccation (drying), and provides a pathway for water and nutrient uptake by bridging the space between the root and surrounding soil particles. A significant proportion of the sugars produced by plant photosynthesis is secreted into the soil as mucigel.

The cell division zone

The root **cell division zone** (Figure 3.6b) is located behind the root cap, and is the source of new cells for root growth. A group of cells between the root cap meristem and the zone of cell division (meristem) of the root is called the quiescent center (Figure 3.5). The cells in this region are also meristematic,

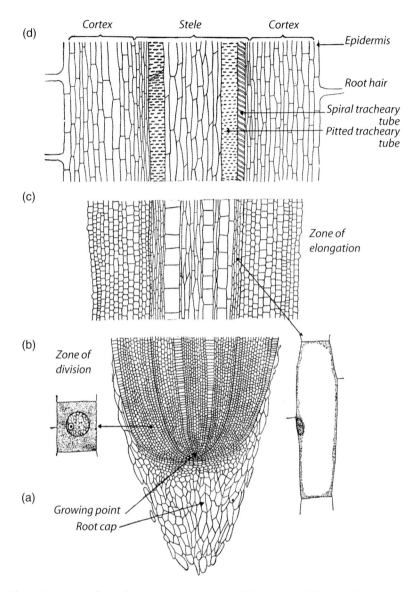

Figure 3.6 Zones of root development: (a) root cap, (b) zone of cell division, (c) zone of cell elongation, (d) zone of maturation. (Drawing from Weaver 1926.)

but divide very slowly, and can provide new cells for both the root and the root cap meristems as needed. Meristems tend to be small; the cell division zone of roots is usually only about 1 mm (0.04 in.) long.

The elongation zone

In the **elongation zone** (Figure 3.6c), cells that are produced by cell division enlarge by taking water into their vacuoles. In roots, which grow mostly in length, cell expansion propels the root tip through the soil as much as 2 cm (0.8 in.) per day. In most roots, the elongation zone extends from the zone of cell division to between 4 and 15 mm (0.015 and 0.6 in.) behind the root tip. It is about 1 cm (0.4 in.) long in most roots, or about 10 times longer than the zone of cell division.

The maturation zone

After cells have divided and elongated, they begin to assume their distinctive roles within the **maturation zone** of the root (Figure 3.6d). Elongation is followed by other modifications that adapt cells for specialized functions, and this process occurs 1–5 cm (0.4–2 in.) behind the root tip. In a study of the primary roots of a grass (barley; *Hordeum vulgare*) and a dicot (marrow or squash; *Cucurbita pepo*), Graham et al. (1974) demonstrated that water uptake is greatest in the region 2–8 cm (0.8–3 in.) behind the root tip in the region of maturation, and is reduced by suberization of the endodermis (Figure 3.7). In the maturation region, root hairs form as extensions of epidermal cells. Root hairs would be sheared off by movement through the soil if they formed in the elongation zone. By greatly increasing root surface

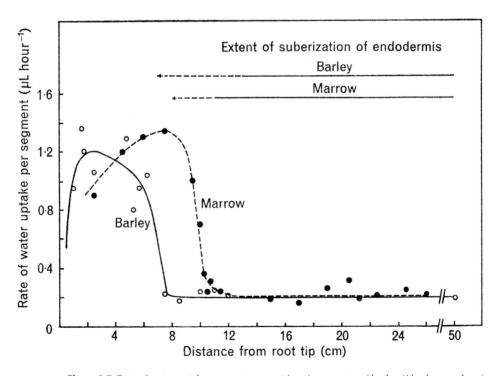

Figure 3.7 Rate of water uptake per root segment in primary roots of barley (*Hordeum vulgare*) and marrow (squash; *Cucurbita pepo*) with distance from the root tip. (Figure from Graham et al. 1974, used with permission of the Biotechnology and Biological Sciences Research Council.)

44 Structure and Function of Plants

Figure 3.8 Root hairs of a maize seedling. (Photograph from Wilkins 1988, used with permission.)

area (Figure 3.8), root hairs increase the absorption potential of the root by several thousand times.

The root dimensions of grasses

In 1937, Dittmer published a study of winter rye (*Secale cereale*) in which he reported the total surface area and length of the shoot and roots of a single plant. In 4 months of growth, the rye plant accumulated 51 ft^2 of leaf surface area (counting both the upper and lower surfaces of leaf blades and the outer surface of leaf sheaths) and a total root surface area of 2,554 ft^2. Dittmer also measured the surface area of root hairs (Figure 3.8) on the same plant; root hairs added 4,321 ft^2 of surface area to the root system. There were a total of 13,815,672 roots, and 14,335,568,288 root hairs, so a single plant had millions of living roots and billions of living root hairs. Dittmer calculated that if the production of living roots had occurred at a steady rate over the life of the plant, 3 mi. of root and 55 mi. of root hair length would have been added to this plant during each day of growth for a total length of 387 mi. of root and an incredible 6,604 mi. of root hair length at 4 months of age.

This single rye plant was grown in a greenhouse in a little less than 2 ft^3 of soil, and root dimensions are reported for the entire plant. When Dittmer measured roots of winter rye grown the field in the top 6 in. of soil (Dittmer 1938), root length was only about 1% that of the greenhouse-grown plant because of competition from other plants, and less than optimal water and nutrient supplies. Given the root dimensions of even field-grown plants, however, it would seem we should see nothing but roots when we dig in the garden. While the fibrous root systems of grasses do contribute large quantities of organic matter to soil, most winter rye roots and root hairs are so thin that in the field study they occupied only 1.275% of the soil volume.

Root tips and their associated root caps are fragile but critically important regions of the root. Understanding their development and function makes it clear why soil adheres to the young roots of plants, and why the root system should be handled carefully during transplanting to retain as much of the growth and uptake functions of the root tips as possible.

Root tissues

The internal structures of the root that can be seen in cross section during primary growth (Figure 3.9) are the epidermis, the cortex, the endodermis, the pericycle, the xylem, and the phloem, and in monocots, the pith.

Epidermis

The epidermis is usually one cell layer thick. Root epidermal cells do not have a thick cuticle, which would interfere with water uptake, and there is no need for stomata because there is no cuticle to limit gas exchange.

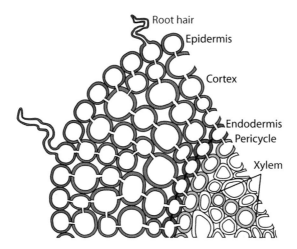

Figure 3.9 Diagram of a root in cross section labeled to identify the tissues.

The cortex

The cortex of the root comprises three layers: the exodermis, the storage parenchyma, and the endodermis.

Exodermis

The **exodermis** (sometimes referred to as the **hypodermis**) is a layer of suberized cells lying just beneath the epidermis (Figure 3.10). The exodermis differentiates in the relatively mature region of the root above (or farther from the root tip than) the maturation (root hair) zone, and is more prominent in roots that have been subjected to drought. The function of the exodermis is to prevent the loss of water that has been absorbed in other parts of the root system to dry soil surrounding mature roots.

Storage parenchyma

Storage parenchyma cells of the cortex are thin-walled cells that can be used by the plant for the storage of starch. Taproots can contain a great deal of storage parenchyma, but in thin, fibrous roots there is only a minimal amount of cortical parenchyma.

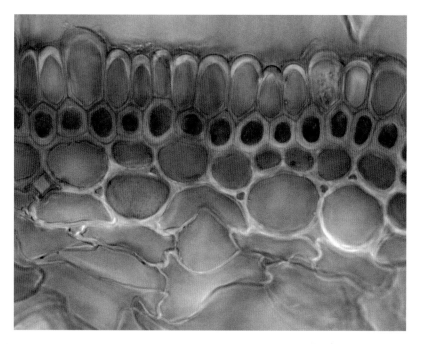

Figure 3.10 Cross section of the outer part of an *Iris siberica* root, 120 mm from the root tip. Stained with Sudan Red 7B for detection of suberin lamellae (red) in exodermal cell walls, and viewed with white light. Tertiary walls are thick and deposited interior to the suberin lamellae. (Courtesy of Chris Meyer.)

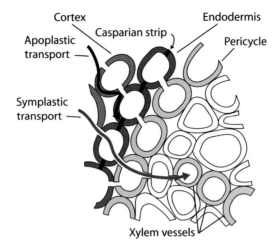

Figure 3.11 Dissolved nutrient ions can move through the apoplast of the cortex, but the Casparian strip of the endodermis blocks movement into the stele. Nutrient ions can move through the symplast into endodermal cells by uptake through the plasma membrane, then into the xylem for transport to the shoot.

Endodermis

The innermost layer of the cortex is the **endodermis**. The walls of these cells are wrapped with a band of suberin and lignin called the **Casparian strip** (Figure 3.11). The function of this layer of cells is to regulate root uptake of dissolved mineral nutrients. The Casparian strip forms around endodermal cells below (or on the root-tip side of) the maturation region, where root hairs form. The Casparian strip blocks the movement of water and dissolved minerals through the walls of endodermal cells, forcing water and dissolved minerals to be taken up through the plasma membrane of the endodermis, cortex, or epidermis before they reach the xylem and phloem tissues of the stele. Outside the endodermis, these substances can move freely through the apoplast (the intercellular space and cell walls), but must enter the symplast (through plasma membranes into living cells) to be absorbed and transported to the shoot (Figure 3.11).

Stele

The **stele** is the collective term for all the tissues located to the inside of the endodermis. These are the pericycle, the vascular tissues, and in some cases the pith.

Pericycle

The **pericycle** is a meristematic tissue one or more cell layers thick that is the origin of branch roots (Figure 3.12). Branch roots form by cell division

48 Structure and Function of Plants

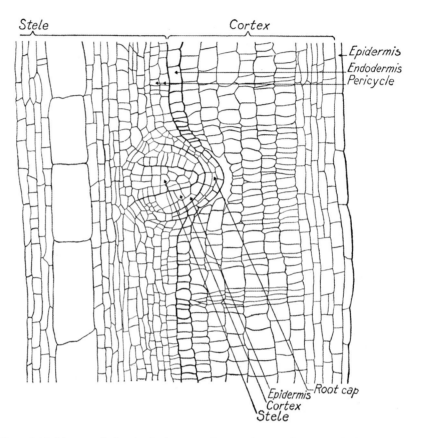

Figure 3.12 Diagram of a longitudinal section of a root, showing the origin of a lateral root in the pericycle. Cellular layers of the new lateral root are labeled according to the mature tissues into which they will develop. (Drawing from Weaver 1926.)

in the pericycle and force their way through the cortex as they develop. The number and location of branch roots is determined by plant hormones.

Xylem and phloem

Root vascular tissue consists of xylem and phloem as in the shoot. The pattern typically formed by xylem cells in dicots is lobed in cross section, with xylem cells in the center of the root and two or more spokes of xylem cells radiating from the center to the perimeter of the stele (Figure 3.13a). The number of lobes or spokes may vary, even within the same plant.

In monocot roots, the pattern of large xylem cells in cross section is cylindrical, with pith tissue in the center (Figure 3.13b). Smaller xylem cells are located between the ring of larger xylem vessels and the perimeter of the stele. Phloem cells in the stele of monocots are located between or outside xylem elements.

Beyond the maturation zone, in the mature root where water absorption is no longer an important function, the endodermis becomes so thickened and

(a) (b)

Figure 3.13 Cross section of the root of a ragwort (*Senecio vernalis*) plant (a). Xylem elements are autofluorescent because of lignification and appear yellow; green fluorescence is the result of labeling to demonstrate the localization of the production of a plant secondary metabolite. (Photograph by Moll *et al.* 2002, used with permission.) The stele of a maize (monocot) root (b) has pith in the center of a ring of large water-conducting metaxylem cells.

waterproof that it creates a seal around the stele. This allows **root pressure** to develop at night, when the demand for water by the shoot has ceased but mineral salt uptake from the previous day continues to draw water into the root in the maturation zone. **Guttation** is seen in grasses as the overnight formation of water droplets at leaf tips (Figure 3.14), and is the result of root pressure forcing water through the stem and leaf xylem to relieve root pressure.

Taproots undergo secondary growth or increase in girth that is similar in the organization of tissue to secondary growth of shoots. In the woody root of hibiscus (Figure 3.15), xylem has formed to the inside of the vascular cambium, and phloem has formed to the outside.

Symbiotic nitrogen fixation

A **symbiotic** (mutually beneficial) relationship can exist between many legume species and rhizobia bacteria. In this relationship, atmospheric nitrogen (N_2) is converted into ammonia (NH_3). The plant receives reduced or "fixed" nitrogen in a form it can use and the bacteria receive carbohydrates and other nutrients from the plant. In their free-living form, rhizobia do

Figure 3.14 Guttation from the tips of orchardgrass (*Dactylis glomerata*) leaves.

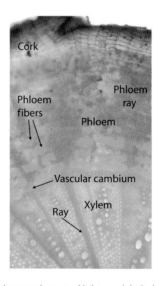

Figure 3.15 Cross section of the woody root of hibiscus labeled to identify the tissues.

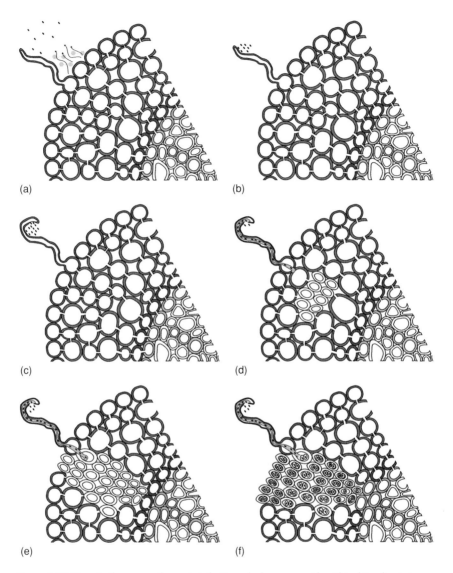

Figure 3.16 Steps in the non-pathogenic infection of a legume root by *Rhizobium* bacteria. The root secretes compounds that attract a specific species of the soil-living bacteria (a). The bacteria attach to the wall of a root hair (b) that breaks down in response to compounds secreted by the bacteria (c). An infection thread containing the bacteria grows back through the root hair cell, while cells in the cortex multiply to create a nodule primordium in anticipation of infection (d). The infection thread reaches the nodule primordium (e) and bacteria enclosed in host cell plasma membrane are released and become bacteroids (f), the form of rhizobia able to fix nitrogen.

not fix nitrogen, because anaerobic conditions are required for **nitrogenase**, the nitrogen-fixing enzyme the bacteria contain, to function. Some woody species form similar relationships with bacteria in the *Frankia* genus.

Rhizobia are free-living soil bacteria that are attracted to roots of legumes by chemical signals (Figure 3.16a). Infection occurs when low soil nitrogen limits plant growth and there is a proper match between a *Rhizobium* species and a legume species. Legumes secrete chemicals recognized by rhizobia in the soil, which migrate to the plant root and secrete their own signal molecules that facilitate their attachment to the cell wall of root hairs (Figure 3.16b).

Root hairs are outgrowths of epidermal cells, and after rhizobia become attached, the wall of the root hair breaks down (Figure 3.16c). An **infection thread**, which is an ingrowth of the epidermal cell plasma membrane, grows back through the cell, enclosing the dividing bacteria (Figure 3.16d). At the same time, cells in the cortex of the root close to a lobe of the xylem begin to multiply in anticipation of infection (Figure 3.16d). The infection thread grows through cortical cells until it reaches the group of newly divided root cells that constitute the nodule primordium (Figure 3.16e). Bacteria, enclosed in host plasma membrane, are released into the nodule cells where they divide further, then enlarge to become **bacteroids**, the form that can fix nitrogen (Figure 3.16f). The **root nodule** develops its own cortex, which limits oxygen

Figure 3.17 Alfalfa (*Medicago sativa*) nodules growing against the transparent PVC face of a rooting container. (Courtesy of Laura Vincent Vaughan.)

diffusion into the nodule, and branches of xylem and phloem develop from the root into the nodule cortex for translocation of water and nutrients.

Nitrogenase, the nitrogen-fixing enzyme, is synthesized by bacteroids but is irreversibly inactivated by high levels of oxygen, which is needed in the same cells for respiration. To keep the oxygen concentration low enough for nitrogenase to function, an oxygen carrier, **leghemoglobin**, is synthesized jointly by bacteroids and the plant. It functions like hemoglobin in blood, carrying oxygen to the enzyme that uses it in the last step of respiration. Like hemoglobin in blood, leghemoglobin with captured oxygen imparts a pink pigmentation to healthy nodules. The pink color of the leghemoglobin can be seen in the healthy, many-lobed nodules of alfalfa in Figure 3.17.

In symbiotic nitrogen fixation, nitrogen gas (N_2) from the atmosphere diffuses into the nodule where it is reduced stepwise by nitrogenase using ATP that comes from respiration (Figure 3.18). In nitrogen fixation, nitrogen gas (N≡N) is successively reduced to HN=NH, H_2N-NH_2 and finally to two molecules of ammonia (NH_3). Ammonia can be toxic to the plant, so it is quickly used to synthesize amino acids in temperate legumes like peas. If legumes are fertilized with inorganic nitrogen such as nitrate (NO_3^-), they will utilize it as readily as nonlegumes and nitrogen fixation will be reduced.

The inoculation process of nodule formation takes up to 4 weeks, so very young seedlings are dependent on seed reserves and soil nitrogen. When the

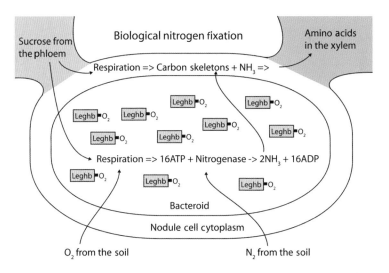

Figure 3.18 In legume root nodules, bacteroid metabolism is fed by carbohydrates from the plant. Bacteroids reduce atmospheric nitrogen (N_2) to ammonia (NH_3) which is used to synthesize amino acids or other nitrogen-containing compounds that can be translocated to the plant. Nitrogenase, the enzyme complex that reduces or "fixes" N_2, is inactivated by oxygen (O_2). However, the fixation process has a high requirement for ATP from respiration, an O_2-requiring process. Therefore, the cytoplasm of nodule cells contains an oxygen-carrying compound, leghemoglobin (also a red pigment), giving active nodules a pink color. Leghemoglobin delivers O_2 to enzymes in respiring bacteroids as it is needed, keeping the concentration sufficiently low for nitrogenase to remain active.

inorganic soil nitrogen is depleted, nitrogen fixation will occur in nodules. Legumes support symbiotic nitrogen fixation for their own benefit, providing from 30 to 95% of the total nitrogen needed by the plant. In meadows, grasses next to legumes are greener, suggesting better nitrogen availability than for grasses growing farther away. There is some evidence for nitrogen transfer between intermingled roots of legumes and other plants, but most transfer from legumes to grasses occurs indirectly from leaf drop or death of the legume plant that releases nitrogen to the soil. Significant redistribution of fixed nitrogen occurs when animals grazing legumes deposit urine and dung near grasses within the same pasture.

Chapter 4

Plant stems

Roots exploit the soil environment in three dimensions, and stems allow leaves to exploit the surface environment in the same way, enabling leaves to be arranged to effectively compete for sunlight and CO_2. Variation in both the internal and external structure of stems is discussed in this chapter, including the outward growth of woody tissues.

The function of stems

1. Stems determine the distribution of the leaf canopy in three dimensions.
2. Stems provide the vascular connections between roots and leaves that allow the transport of water and mineral nutrients from the soil to the shoot, and the distribution of sugars synthesized in leaves ("sources") to nonphotosynthetic tissues or regions of rapid growth that need energy and building blocks for new tissues ("sinks") or to regions of storage that may be located in roots, stems, fruits, or seeds.

The structure of stems

Stems consist of **nodes**, where leaves and **axillary buds** are attached, and **internodes**, which are the regions between nodes that elongate to give height to a stem (Figure 4.1). The shoots of some plants appear to have no stem, while the shoots of other plants appear to have nothing but a stem. Plants that have a **rosette** habit of stem growth have unelongated stem internodes, and their leaves, while attached at different nodes of the stem, appear to originate at almost the same location (Figure 4.2). The leaves of cacti (Figure 4.3), which are specialized for survival in dry, hot climates, are reduced to spines (Figure 4.3 inset), and cactus stems are the site of both photosynthesis and water storage.

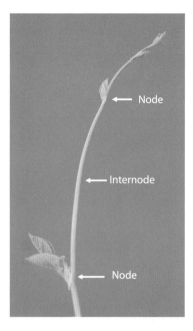

Figure 4.1 The shoot tip of a bean (*Phaseolus*) plant with an elongated internode bracketed by nodes, at which leaves are developing.

Figure 4.2 The leaves of this dandelion (*Taraxacum*) are arranged in a rosette. In lawns, dandelion leaves compete effectively with surrounding grasses for light.

Figure 4.3 The body of this cactus consists mostly of the stem. Leaves have been reduced to highly lignified masses of fibers, or spines, which provide effective protection from herbivores.

Axillary buds are meristematic regions that form in the axil (or junction) of the stem and a leaf and develop into inflorescences or branches. Stolons, rhizomes, and tillers are all formed from axillary buds and produce independent plants.

Stolons (Figure 4.4a) are aboveground horizontal stems or branches that explore the soil surface to find nutrient-rich microsites. Stolons can be seen in strawberry (*Fragaria*), spider plant (*Chlorophytum comosum*), buffalo grass (*Bouteloua dactyloides*), and white clover (*Trifolium repens*). Stolons have nodes and internodes just like other stems, and where the nodes come into contact with soil, they can form roots and a new shoot. Thus, stolons are a means of vegetative propagation of these plants. Because the new plants form from the mother plant without sexual reproduction, they are genetically identical to the original plant.

Figure 4.4 (a) A stolon of Bermuda grass (*Cynodon*) with shoots developing from axillary buds at nodes. If these nodes encounter soil, they will also develop roots. (b) These rhizomes of a switchgrass (*Panicum virgatum*) plant grown in a 1-gallon pot are searching in all directions for soil to explore. (c) A single plant of a bunchgrass with myriad tillers and subtillers, all developing from axillary buds on a central, unelongated stem. Each new leaf is associated with an axillary bud that can produce a tiller.

Rhizomes (Figure 4.4b) are underground horizontal stems that also serve as storage for food reserves. Like stolons, rhizomes have nodes and internodes. They root at nodes and send up shoots. A number of problematic weeds like Canada thistle (*Cirsium arvense*) and quack grass (*Elymus repens*) spread by rhizomes. Cultivation (e.g., tilling) of these weeds breaks up and spreads the rhizomes and actually increases plant numbers, one of the reasons these weeds are hard to control.

Tillers originate from axillary buds, usually at the unelongated nodes of grass stems (Figure 2.1). Grasses are usually classified as bunchgrasses or sodforming grasses. Sodforming grasses like Kentucky bluegrass (*Poa pratensis*) produce rhizomes from their axillary buds and fill the space between established plants with new plants creating a smooth, uniform turf. Bunchgrasses

(a) (b)

Figure 4.5 (a) The twining shoot of a bindweed (*Convolvulus*) vine climbing a fence. (b) A tendril from the stem of this Virginia creeper (*Parthenocissus quinquefolia*) supports the weight of many large leaves.

(Figure 4.4c) form new tillers from axillary buds on the main stem. The axillary buds on those new tillers produce their own tillers, and so on. This results in a clump of grasses in which the central plant may die, leaving a ring of later generations of the same plant.

Twining shoots (Figure 4.5a) are modified stems, and **tendrils** (Figure 4.5b) are leaves or branches modified to provide metabolically inexpensive support. Unequal growth of these organs causes motion in the stem tip that eventually brings them into contact with external support. After contact, the shoot or tendril grows in a tight spiral around the support, allowing the plant to use new carbohydrates from photosynthesis for further growth rather than investing in the secondary walls of fiber cells for upright support.

Bulbs such as onion, lily (Figure 4.6a), tulip, and daffodil have compact unelongated stems with fleshy underground leaves that surround the apical meristem and store food. The apical meristem becomes the flower stalk and leaves, and the axillary buds of the bulb's fleshy leaves develop into the next year's bulbs. **Corms** (Figure 4.6b) are thickened underground stems consisting of several nodes and unelongated internodes that increased in diameter to store food reserves. When corms are planted, the uppermost axillary bud will break dormancy and elongate into a flower stalk fed by the mother corm. Crocus and gladiolus "bulbs" are corms. Dahlias (Figure 4.6c) grow from tuberous roots, which are modified stem tissue with swollen

Figure 4.6 The bulbs formed from a lily (a), the developing daughter corm and older mother corm of a gladiolus (b), and many tuberous roots of a dahlia plant (c) late in the growing season.

regions where food is stored for later growth. As in the case of potatoes, each **tuber** must have at least one bud for a new plant to form.

For bulbs, corms, rhizomes such as bearded iris, tubers such as begonia, and tuberous roots, an ornamental flower stalk and leaves grow from either the apical meristem or an axillary bud. The development of the flower and new leaves is supported by carbohydrates stored in these fleshy underground storage organs. After flowering, carbohydrates produced by the leaves are stored for the following season in the form of new bulbs, corms, rhizomes, tubers, and roots.

Internal tissues of the stem

Stems consist of an epidermis, cortex, vascular bundles (xylem and phloem), and pith (Figure 4.7). In most dicot stems, the vascular bundles are organized in a ring in the cortex that encloses the pith. In the immature stem of maize, a monocot, small vascular bundles are scattered within the ground tissue (Figure 4.8). In more mature monocot stems, the pith may deteriorate,

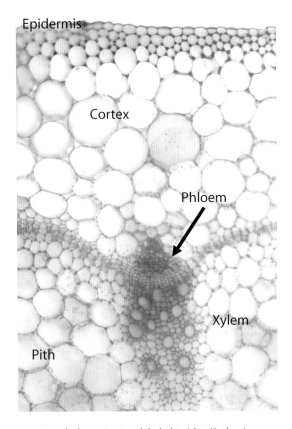

Figure 4.7 A cross section of a begonia stem labeled to identify the tissues.

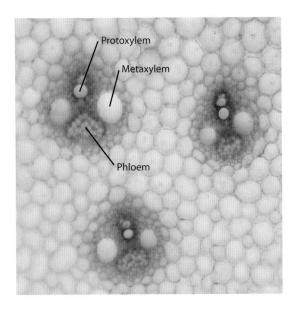

Figure 4.8 A cross section of an immature maize (*Zea mays*) stem.

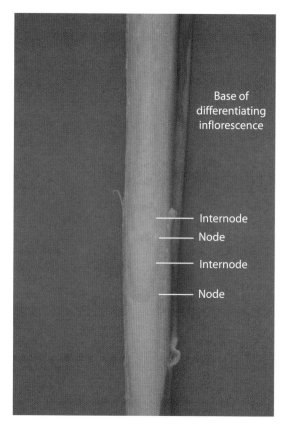

Figure 4.9 The nodes and internodes of an elongating maize stem. The apical meristem (Figure 2.1) has developed into the tassel (male inflorescence) and is being pushed through the whorl of leaf sheaths to the top of the plant.

leaving a hollow stem. The function of mature grass stems is to extend the inflorescence that has developed from the apical meristem (Figures 2.1 and 4.9) above the canopy by the growth of internodes to facilitate pollination. These long, thin stems have many layers of fiber cells interspersed with vascular bundles and islands of cortical cells close to the epidermis (Figure 4.10).

Secondary or woody growth of stems

Growth in height or length is considered primary growth, while increase in circumference or growth is considered secondary growth. Both primary and secondary growths occur in woody perennial roots and shoots. While apical meristems of the roots and shoot are the source of primary growth such as in Figure 4.11, the vascular cambium adds new vascular tissue (xylem and phloem) and the cork cambium adds new layers of protection to the outside of stems each year.

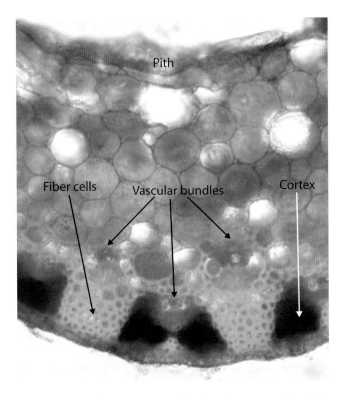

Figure 4.10 A cross section of the mature stem of a wheat (*Triticum aestivum*) plant.

Figure 4.11 The new primary growth (in length) of these maple (*Acer*) branches is in red, while older stem sections that have undergone secondary growth are lighter in color.

Secondary vascular tissues

Secondary xylem, the vascular tissue that transports water and mineral nutrients up the stem in woody plants, and **secondary phloem**, the tissue that transports sugars and other organic compounds along the outside of the stem, are produced by the vascular cambium. The xylem and phloem of a young linden (*Tilia*) stem can be seen in Figure 4.12. The xylem has an abundance of water-carrying cells, while the phloem is filled with several rows of nearly unstained phloem fibers. Phloem rays fill the space between groups of phloem cells that develops as the circumference of the stem increases.

The tissue formed to transport water and nutrients between the inside and outside of the stem (radially) is formed of ray parenchyma cells. In a longitudinal section of pine (Figure 4.13), the long tapering (fusiform-shaped) tracheary water-conducting cells are interspersed with columns of small, round ray cells.

Division of **fusiform initials** is through the long axis of the cell. This differs from division in other meristems, where division is across the long axis of the cell and which adds height or length. A stem increases its circumference by way of longitudinal divisions that create two cells from each fusiform initial. A new cell that forms to the inside of the vascular cambium becomes a xylem (wood) cell, while a new cell that forms to the outside of the vascular cambium becomes a phloem (bark) cell. Initials in the vascular cambium divide over and over to produce these new vascular tissues. The vascular cambium produces several times more xylem than phloem cells, and the result is that wood comprises about 90% of a typical tree trunk. In winter, the vascular cambium becomes dormant (inactive).

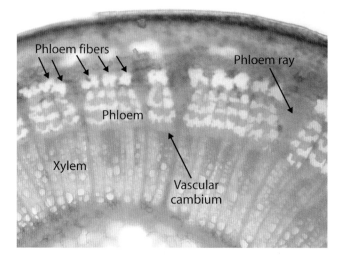

Figure 4.12 A cross section of a young linden (*Tilia*) branch.

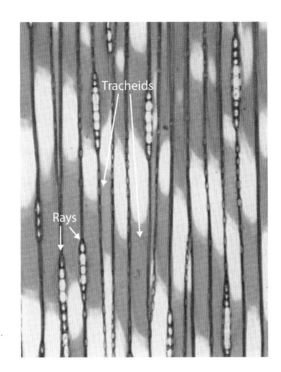

Figure 4.13 A longitudinal section through pine wood composed of tracheids (water-carrying cells) and rays. (Photograph from Wilkins 1988, used with permission.)

The secondary xylem produced in the spring (**springwood**) consists of large, relatively thin-walled cells while the wood produced in summer (**summerwood**) consists of smaller, thicker-walled cells. Cell growth takes place through water uptake, which forces the growing cell wall to expand. In spring, there is a greater need for transport of water for growth of new leaves and branches of the shoot. Less water is needed in the summer after leaf growth has stopped. **Annual rings** that are visible in cross sections of tree trunks from temperate regions are the result of change in cell size from large to small over the course of each growing season. In drought years, narrow tree rings are the result of fewer cell divisions in the vascular cambium.

Woods are classified as softwoods or hardwoods. **Softwoods** are from conifers growing in temperate environments, and this wood contains only tracheids and lacks vessels (Figure 4.14a). Softwoods are lightweight and their tracheid cell walls are high in lignin, which makes them more resistant to warping. Softwoods are relatively easy to nail, and are therefore ideal for lumber. Cured or dried lumber contains about 15% water, 65% cellulose, and 20% lignin by weight.

Hardwoods are from dicots like maple, and contain thick-walled fibers and large-diameter vessels in addition to tracheids. Fibers increase the density

66 Structure and Function of Plants

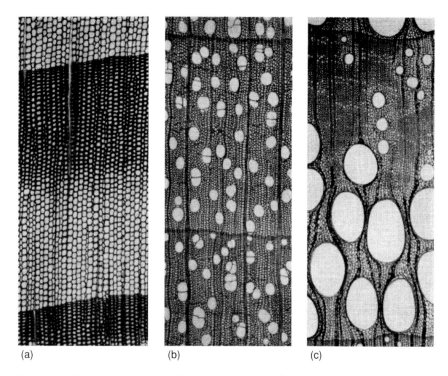

Figure 4.14 Cross sections of wood from the primitive conifer pine (*Pinus*) in which water is conducted by tracheids (a), birch (*Betula*) a tree at an intermediate stage of evolution with diffuse-porous wood (b), and oak (*Quercus*) an advanced tree species with large-porous wood (c). (Photograph from Zimmermann 1963, used with permission.)

and strength of wood, and hardwoods are utilized for furniture construction. In some like birch (*Betula*; Figure 4.14b), the water-conducting vessels are smaller in diameter and still contain end walls that are not completely dissolved. Advanced hardwoods such as oak (*Quercus*) (Figure 4.14c) have large-diameter xylem elements without end walls. These hardwoods may also have parenchyma cells that can absorb water from the water-carrying vessels and tracheids at night or after a rainfall, and store it to be used in dry, hot weather. Parenchyma cells are also the source of the tannins and resins that fill the cells of heartwood.

Figure 4.15 shows a wedge from the trunk of a hardwood tree. The outermost layers of the bark are nonliving (A) but the inner bark (phloem) transports dissolved organic compounds. The boundary between layers B and C is the vascular cambium, which divides to produce both the xylem and phloem. The layers marked C comprise three annual rings of water-conducting **sapwood**, and D comprises seven annual rings of heartwood. As trees age, the central xylem becomes a repository for gums, resins, oils, and tannins, which plug the xylary elements. This dark-stained **heartwood** no longer carries water.

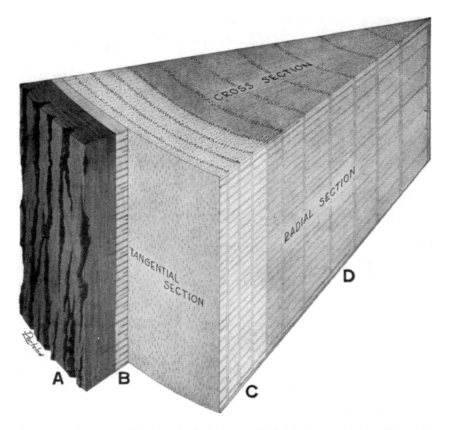

Figure 4.15 The tissues of hardwood include the outer bark (A), the inner bark or phloem (B), the functioning sapwood (C), and the heartwood (D) which no longer carries water. (Figure from Brown et al. 1949, used with permission of the McGraw-Hill Companies.)

Bark—secondary phloem and cork

The **bark** is considered to be all the tissues outside the vascular cambium and therefore includes both the phloem and the cork. Secondary phloem conducts water and organic solutes between the leaves and roots of a tree. Like the wood (xylem), the secondary phloem is derived from the vascular cambium, and includes sieve elements, parenchyma storage cells, and bands of thick-walled fibers. Remember that the primary xylem and primary phloem are the earliest-formed stem tissues. The primary xylem will therefore be the oldest xylem, and will be located in the center of the tree trunk farthest from the vascular cambium, while the primary phloem, also the first formed and oldest, will be located to the outside, closest to the epidermis, and farthest from the vascular cambium.

The functional sieve elements of the secondary phloem live less than 1 year, and are those within about 1 cm of the vascular cambium; those farther to

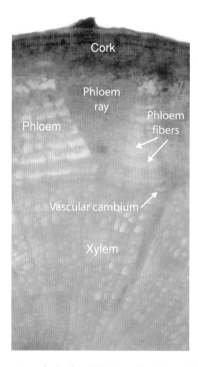

Figure 4.16 In this cross section of a linden (*Tilia*) branch, phloem fibers (green) are interspersed with functioning phloem (gold), and wedge-shaped phloem rays fill the space created by secondary growth in circumference. The youngest xylem and phloem cells are on the inside and outside, respectively, of the vascular cambium.

the outside are dead (and crushed) and are therefore nonfunctional. Only the fiber cells, which retain their shape because of their lignified cell walls, can still be seen in much of the phloem. While new xylem cells are added to the out side of older xylem cells, the phloem grows on its inside face. As a tree trunk grows in circumference, and the oldest phloem is pushed farther from the center of the trunk, rays fill in the space between other tissues (Figure 4.16).

In order to maintain an outer covering on a tree trunk that is continuing to increase in diameter, **cork** is formed. It is an outer suberized layer that replaces the epidermis, and which protects and insulates the underlying tissues (Figure 4.17). This layer must be continuously replaced during the growing season and therefore requires its own meristem, called the **cork cambium**. The cork includes one or more layers of cork cells stacked to the outside of the cork cambium (Figure 4.18), and may also produce a secondary cortex to the inside of the cork cambium. The outer bark is considered to be all the tissues from the outside of the cork cambium, while the inner bark is made up of the tissues from the vascular cambium to the cork cambium.

There are pores in the outer surface of bark called **lenticels** that act as a conduit for gas exchange across the cork. Lenticels are the dark lines on corks. These channels are necessary because the phloem is very active metabolically, requiring O_2 for respiration and expiring CO_2.

Figure 4.17 The bark of a sycamore (*Platanus*) tree.

Figure 4.18 Cork cells of a hibiscus root.

Commercial cork production

Commercial cork, such as that used to seal wine bottles, comes from the cork oak (*Quercus suber*) tree, an evergreen oak native to the western Mediterranean, with about half the world's production coming from Portugal. In one season, the cork cambium may produce as many as 40 layers of cork cells. It takes about 25 years for the cork oak to accumulate a first harvestable crop of cork, a layer 3–10 cm (1–4 in.) thick. Successive harvests can occur every 6–10 years after the first, and cork oak trees remain productive for 150 years or more. The cork layer of these trees peels away at the cork cambium layer. Because removal of the cork leaves the vascular cambium and the phloem intact, cork oak trees are not harmed by the removal of the cork layer and the cork is regenerated.

The cells of cork tissue are heavily suberized and dead at maturity. **Suberin** is a waxy substance that makes the cork cells waterproof and airtight. Because of the suberization of the walls of its cells, cork prevents the movement of both liquid and gas across its cells. Commercial cork is exceedingly light, because half of its volume is trapped air, so it is used in fishing for floats and buoys. Cork is also exceedingly resilient to compression. The joints at the segments of woodwind instruments such as clarinets are lined with cork, making the seals airtight when the instrument is assembled. The corks that are used to seal wine bottles are cut such that their lenticels run perpendicular to the neck of the wine bottle. In Figure 4.19, the lenticels of a number of corks can be seen as dark streaks across the cut ends.

Figure 4.19 These corks from cork oak (*Quercus suber*) were cut such that the lenticels, the channels that provided gas exchange to the inner bark, run across the cork rather than up and down through the tissue.

Chapter 5

Plant leaves and translocation

Leaves vary in size, shape, and complexity. Leaves from some plants in temperate climates are created to last only one season, but before their skeletons fall to the ground in autumn, their nutrients are recycled; this process causes leaves to change color in the autumn. Other leaves invest a great deal to make each leaf durable so they can last many years and are considered "evergreen." The structure of a variety of leaf types is discussed in this chapter, while the main function of leaves—photosynthesis—is explored in Chapter 10.

The function of leaves

By far the most important function of leaves is photosynthesis. Photosynthesis is the first link in the food chain for all living creatures, creating hundreds of billions of tons of sugar each year. Most leaves are constructed as a trade-off between optimal surface area for light interception and protection from excess water loss. Photosynthesis requires uptake of carbon dioxide (CO_2) from the air and produces oxygen (O_2), so leaves need to have extensive internal air spaces. They utilize the stomata (pores) in the epidermis for gas exchange, and leaves in challenging environments may develop channels within the mesophyll for internal gas exchange.

The leaves of dicots tend to be oriented horizontally, while grass leaves tend to be more vertical in orientation. This is why clovers and grasses can coexist well in lawns—the blades of grass can fit between the flat clover leaves and still harvest the light they need for their own growth (Figure 5.1).

The structure of leaves

Leaf blades may be **simple**—one undivided blade (Figure 5.2a)—or **compound**, where each leaf consists of several leaflets (Figure 5.2b). The

Figure 5.1 The flat, round leaflets of white clover (*Trifolium repens*) growing among the narrow, vertical leaves of Kentucky bluegrass (*Poa pratensis*) in a lawn.

arrangement of leaflets of a compound leaf may be **palmate**, where all the leaflets are attached to the petiole at a single point, as in the true clovers (Figure 5.2b), or **pinnate** (feather-like), where leaflets are arranged in pairs along the leaf axis, as in these sainfoin leaves (Figure 5.2c). There are many other variations of compound leaves, including subdivided compound leaflets (Figure 5.2d).

Leaves are attached to stems at nodes, and there may be only one or dozens of leaves arranged around a single node. Leaves may be attached directly to the stem ("**sessile**") as in this penstemon (Figure 5.3) or attached by a **petiole**, which is a stalk attached to the stem at a node (Figure 5.2b). Grasses have sessile leaves (Figure 5.4), consisting of a vertical sheath that encloses either the stem or less-developed leaves, while the blade is relatively horizontal and more active in light interception. The petioles of a **rosette** plant (growing close to the ground because stem internodes are unelongated) may differ in length to orient leaves, so they do not overlap to maximize light interception

Figure 5.2 Variations in leaf blades: (a) a maple (*Acer*) leaf consists of a single blade, (b) the true clovers (*Trifolium*) have compound leaves composed of three leaflets. When each leaflet is attached to the same point on the petiole it is palmately compound; when the leaflets are attached at different points along a central axis (c) the leaf is pinnately compound. The arrangement of leaflets on the prairie bundleflower (*Desmanthus illinoensis*) leaf (d) is similar to the sainfoin (*Onobrychis viciifolia*) leaf, but each leaflet is subdivided.

(Figure 5.5). There is usually one axillary bud associated with every leaf, located just above the site of attachment of the leaf to the stem. A pair of **stipules**, scale- or leaf-like appendages that can enclose the developing leaf and axillary bud, may also occur at the axil of the leaf and the stem (Figure 5.6).

Leaves use a variety of mechanisms to adapt to their environments. Some leaves use **pubescence** (hairiness) to reflect light and repel insects. With heavy pubescence, leaves may appear almost silver (Figure 5.7). Pubescence also decreases the movement of air next to the leaf, reducing the rate of transpiration. In very harsh environments, plants can reduce transpiration significantly by decreasing their surface-to-volume ratio. The sphere is the

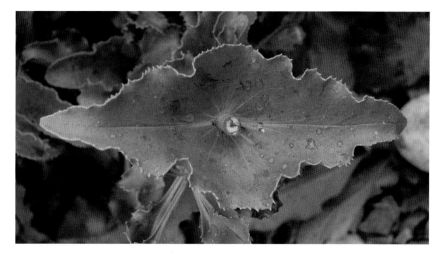

Figure 5.3 This penstemon leaf is sessile, attached directly to the stem without a petiole.

three-dimensional shape with the lowest surface-to-volume ratio, so the thicker and more spherical a leaf, the less surface there is for transpiration (Figure 5.8). Some succulents may even grow below ground, with only their leaf surface exposed to the sun. This not only reduces air movement around the leaves, but the soil insulates the leaves, reducing daytime temperatures and modulating the change from high daytime temperatures to the low

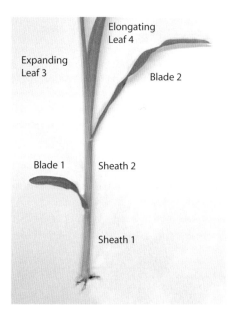

Figure 5.4 Grass leaves consist of an upright sheath that encloses younger leaves and eventually the stem, and blades that are upright as they elongate through the whorl of older leaves, then more horizontal when fully expanded.

Figure 5.5 The petioles of this arrowleaf balsamroot (*Balsamorhiza sagittata*) help arrange each leaf to intercept sunlight with minimal interference by other leaves. (Courtesy of James Sinclair.)

Figure 5.6 Stipules are the leaflike appendages that occur at each node of the stem of this pea seedling. The function of stipules in enclosing the developing leaf at young nodes can be seen in Figure 4.1.

Figure 5.7 Leaves of lambs ears (*Stachys byzantina*) are thickly pubescent. The hairs on the leaf reflect light, create a barrier to insect pests, and slow the loss of water from the leaf by creating a boundary layer (still air) near the leaf surface.

Figure 5.8 The leaves of this succulent (*Crassula*) are thick relative to their width, minimizing their surface-to-volume ratio to reduce leaf surface area for transpiration.

(a)   (b)

Figure 5.9 Bud scales of a black maple (*Acer nigrum*) opening in the spring (a) to reveal flowers and new leaves (b).

nighttime temperatures found in desert environments. Cacti are desert plants that have reduced many of their leaves to **spines** to minimize transpiration (Figure 4.3).

Modified leaves also protect the apical meristems of the branches of woody perennials. **Bud scales** are thick, leathery leaves that enclose apical and axillary meristems, protecting them from freezing and dehydration during the winter (Figure 5.9a). In spring, the bud scales open and meristems resume growth (Figure 5.9b).

In low-nitrogen environments, plants may use specially modified **trapping leaves** to capture insects for nitrogen. Such leaves may be passive traps, where the insect cannot retrace his path because it is blocked with downward-pointing spines (Figure 5.10a), or active traps that close over the insect in response to mechanical stimulation (Figure 5.10b). In both cases, trichomes on the epidermis that lines the trap secrete digestive fluids, and the epidermis absorbs the nutrients.

Tendrils are modified leaflets adapted to grow around vertical supports encountered by the plant (Figure 5.11). In this way, the plant does not need to invest as much carbohydrate in the secondary walls of the fiber cells that would normally provide the structural support for stems and leaves. The opposite strategy is employed by plants with **sclerophyllous** leaves, such as yucca (Figure 5.12), that contain more carbohydrate in their highly fibrous leaves than the leaf could justify for photosynthesis in 1 year. These leaves are so durable that they are perennial, and the plants are said to be "evergreen." The fibers of yucca were used by the Ancestral Puebloan people

Figure 5.10 The epidermis lining the pitcher plant (*Nepenthes*) reservoir (a) and the inner surfaces of the venus flytrap (*Dionaea*) lobes (b) secrete enzymes that digest captured insects, releasing nitrogen to the plant. (Photographs from Wilkins 1988, used with permission.)

in the American Southwest to make baskets, rope, and moccasins. Conifer needles (Figure 5.13) and cedar scales are also leaves modified to be sufficiently drought-tolerant and winter-hardy to last for several years. Conifer needles have a hypodermis inside the epidermis to reduce transpiration and an endodermis that encloses the vascular tissue, and stomata are located in depressions in the leaf surface.

The development of leaves

Leaves are initiated at the apical meristems of stems or branches as leaf primordia. As they develop, each new **leaf primordium** surrounds and protects the apical meristem. In developing monocot leaves, meristems are located at the base of the leaf. Monocot leaf cells divide in these basal meristems and then expand and differentiate linearly, like roots (Figure 5.14). Division in very young dicot leaves occurs throughout the leaf blade in two dimensions (length and width). Dicot leaves mature first at the tip and margins, with

Figure 5.11 Some leaflets of the compound leaves of peas are tendrils that can provide vertical support in place of fiber cells.

Figure 5.12 A yucca plant with fibers curling along the margins of the leaf.

Figure 5.13 The needles of conifers are compact and cylindrical to minimize water loss while exposing all aspects of the needles to sunlight.

divisions continuing near the leaf base until the leaf is one-half to three-fourths its mature size (Figure 5.15).

The **mesophyll** of leaves includes both thin-walled photosynthetic cells and supporting fiber cells. The photosynthetic cells of monocot leaves are arranged in layers around veins (Figure 2.12). In dicots, where leaves are

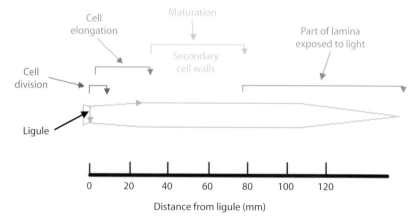

Figure 5.14 The regions of monocot leaf development (cell division, cell elongation, and maturation) are arranged linearly as in developing roots (Figure 3.6), but in the case of grass leaves, the tip is the most mature tissue, while the meristem is located closest to the stem. (Courtesy of François Gastal.)

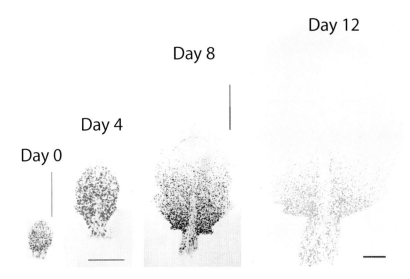

Figure 5.15 Cell division in leaves of *Arabidopsis* from the beginning of its growth, apparent as blue staining from β-glucuronidase (GUS) reporter expression associated with a protein active during cell division. On day 0, cell division is uniform throughout the leaf. On day 4, cell division is reduced at the leaf tip and at the base of the midvein. By day 8, cell division has stopped in the upper and lower epidermis of the distal half of the leaf, with scattered divisions continuing in the mesophyll. By day 12, halfway through this leaf's growth period, cell divisions are restricted to the basal one-third of the leaf blade. Growth by cell expansion continues until day 24. The scale in the photographs differs: at day 0 it is 0.25 mm; at day 4, 0.50 mm; and at days 8 and 12, 1 mm, so leaf dimensions differ far more than suggested by the photographs. (Figures from Donnelly et al. 1999, used with permission.)

oriented horizontally and light intensity is greater in upper leaf tissues, the **palisade** (upper) **parenchyma** cells are elongated and closely packed (Figure 5.16). Palisade parenchyma perform up to 90% of the photosynthesis in dicot leaves. The **spongy** (lower) **parenchyma** cells of dicot leaves are not only connected to each other but also incorporate extensive air spaces.

Leaf veins are located in the center of the leaf between the palisade parenchyma and spongy mesophyll. Within the vein, the xylem (for delivering

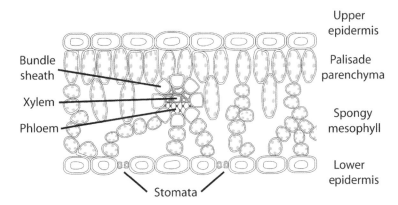

Figure 5.16 A diagram of the anatomy of a dicot leaf labeled to identify the tissues.

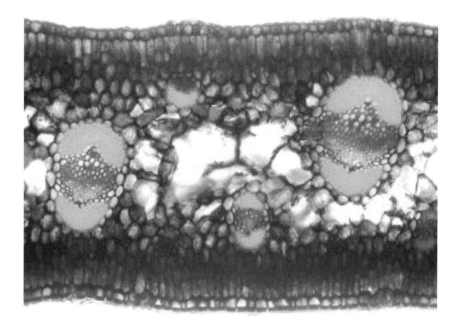

Figure 5.17 The internal anatomy of a leaf of yucca with densely packed, elongated mesophyll cells under both the upper and lower epidermis. Vascular bundles are surrounded by blue-stained fiber caps.

water) is always on the upper side and the phloem (for removing the sugars produced by photosynthesis) is always on the lower side (Figure 5.16). The orientation occurs naturally as veins branch outward from the stem, where the phloem is on the outside (part of the bark) and the xylem is on the inside (the wood). In some leaves adapted to intense light, the mesophyll may be equally dense under both the upper and lower epidermis (Figure 5.17).

The arrangement of major veins in leaves may be **reticulate** (like a net), which is most typical of dicots (Figure 5.18a), or parallel, which is typical of monocots such as grasses (Figure 5.18b). In dicots, each small vein ending is the site for exchange of water and solutes with a few dozen surrounding mesophyll cells. In monocot leaves, minor veins run parallel to the major veins, but there are many lateral connections between minor veins that are perpendicular to the **parallel venation**. In both dicots and monocots, most photosynthetic cells are only one or two cells removed from a minor vein.

Environmental effects on leaf development

Light (sun versus shade)

The amount of light leaves receive will influence the internal structure of different leaves on a single plant. "Sun" leaves are smaller in area and thicker

(a) (b)

Figure 5.18 The reticulate or netted venation of a dicot (a) and the parallel venation of a monocot (b) leaf. (Photograph (a) from Wilkins 1988, used with permission.)

(Figure 5.19a) than "shade" leaves (Figure 5.19b) with more layers of mesophyll under a unit of leaf surface area than **shade leaves**. This is because **sun leaves** receive higher intensity light over a unit of leaf surface area compared with shade leaves, and change their structure to best exploit the light environment. Leaves developing in shade will take the same number of cells and

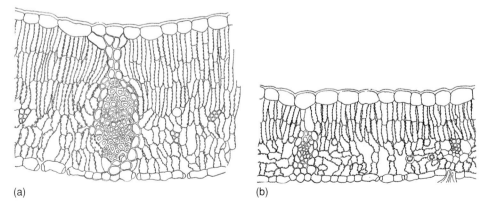

(a) (b)

Figure 5.19 The internal anatomy of a sun (a) and a shade (b) leaf of gambel oak (*Quercus gambelii*). The local intensity of light can influence the anatomy of individual leaves on a single plant. (Drawings from Pool 1923.)

Water

Xerophytes are plants that have become adapted to grow in arid regions or to survive prolonged drought. They often have small, thick leaves (they are extreme versions of "sun" leaves) that have few air spaces and are modified to store water (succulents). Their epidermis may be several cell layers thick (Figure 5.20), with thick cell walls and thick cuticles. Stomata may only occur on the underside of xerophyte leaves, and be sunken and surrounded by hairs (trichomes) to reduce air movement and therefore transpiration. Some desert plants open their stomata only at night.

Hydrophytes are plants that are adapted to wet environments, such as the aquatic plants used in aquariums (Figure 5.21). Plants that live on or under water must also be adapted for reduced light and reduced gas exchange. Stomata may be absent in submerged leaves, but tissue containing large

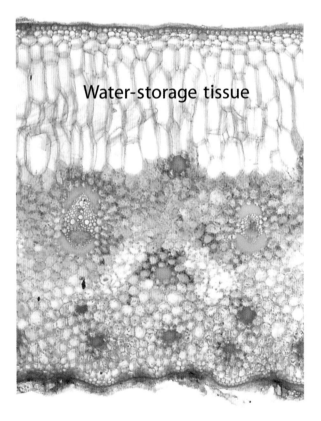

Figure 5.20 The internal tissues of a leaf of pineapple (*Ananas comosus*), a xerophyte with thick cuticles and several layers of water-storage tissue below the upper epidermis.

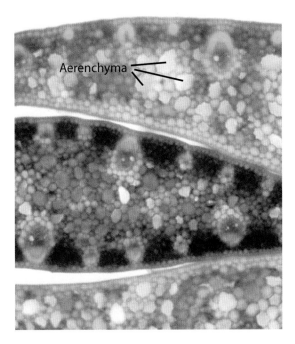

Figure 5.21 The internal tissues of leaves of sweetflag (*Acorus*), a hydrophyte with thin cuticles and many aerenchyma (airspaces) among the ground cells of the leaf.

continuous internal air spaces called **aerenchyma** is common in hydrophytes, and provides the needed pathways for gas exchange. Since water loss is not a problem, hydrophytes usually have relatively thin epidermal cell walls and cuticles.

Translocation

Translocation is movement through the vascular system from one part of the plant to another. Sucrose is made from the products of photosynthesis in chloroplast-containing (green) cells of the plant, most commonly in mesophyll cells in leaves. When a leaf is mature enough to produce more **photosynthate** (the sugars produced by photosynthesis) than the leaf requires for its own growth and maintenance, it becomes a "**source**" leaf.

> **The transition from sink to source**
>
> In Figure 5.22, the fifth leaf from four different squash (*Cucurbita pepo*) plants is shown at different stages of development. The older Leaf 3 on each of the same plants was fed radioactively labeled carbon dioxide ($^{14}CO_2$) 2 hours before Leaf 5 was used to expose X-ray film. Wherever carbohydrates imported from Leaf 3 were

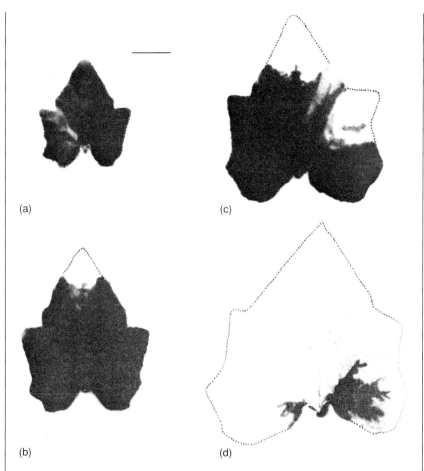

Figure 5.22 In squash (*Cucurbita pepo*), dark regions indicate the incorporation in Leaf 5 of carbohydrates imported from the older Leaf 3. In (a), Leaf 5 is 10% expanded and 85% labeled; in (b), it is 15% expanded and 70% labeled; in (c), 25% expanded and 45% labeled; and in (d), 45% expanded and 8% labeled. All leaves at the same magnification; bar = 2 cm. (Figure from Turgeon and Webb 1973, used with permission.)

used for new cell walls in Leaf 5, radioactivity exposed the X-ray film and created a dark image. The $^{14}CO_2$ from Leaf 3 was used in photosynthesis (Chapter 10) to make sugars. The labeled sugar was exported from Leaf 3 as sucrose, and was translocated through the phloem to the most demanding nearby sinks. Dark shading indicates the localized incorporation of radioactive carbon into the structure of Leaf 5. At 10% expansion (5.22a), about 85% of Leaf 5 was labeled, indicating most parts of the leaf were still forming and using photosynthate produced in Leaf 3. Therefore, at this age, Leaf 5 is a "sink," still requiring import of carbohydrates to support its growth. The leaf finishes its growth first at the margins of the leaf, beginning at the tip (5.22b) then at the lateral margins (5.22c), with dark regions that indicate incorporation of labeled CO_2 retreating down the midvein toward the petiole. The oldest and therefore the largest leaf in this study (5.22d) is able to produce more photosynthate because of its size (45% expanded) and needs very little imported carbohydrate (only 8% of this leaf is labeled), so this leaf is or will soon become a "source" of new carbohydrates for the rest of the plant, exporting newly formed carbohydrates to the most demanding nearby sinks.

The taproots of perennial plants are sinks for carbohydrates, especially in autumn in temperate climates when plants must become winter-hardy if they are to survive. Carbohydrate storage accounts for as much as 40% of root dry weight of alfalfa (*Medicago sativa*), a perennial herbaceous plant, in autumn. In the spring, however, when the previous season's leaves are dead from exposure to below-freezing air temperatures, the taproot and belowground crown of the plant becomes a source of carbohydrates, supporting the development of new stems and leaves from axillary buds. When such perennial plants encounter stress during the growing season that interferes with the storage of carbohydrates before fall dormancy, whether the stress is from drought, nutrient deficiency, insects, pathogens, or other sources, they may not regrow the following spring. While it may appear that the plant died from the cold—since it was alive the previous autumn—the real cause is the stress that prevented the accumulation of carbohydrates.

Excess photosynthate is moved from mesophyll cells to the bundle sheath surrounding a vascular bundle in the leaf and then is loaded into the phloem for long-distance transport to parts of the plant where it is needed (Figure 5.23). This photosynthate is translocated to rapidly growing parts of the plant, such as new leaves and stems, or to plant organs that are not photosynthetically active, such as roots or developing fruit or grain. Anywhere

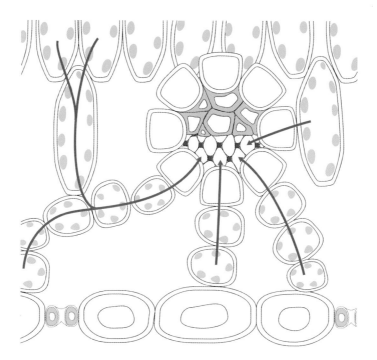

Figure 5.23 The path of movement of sucrose from source leaf cells to the phloem for transport to sinks. Movement may be both symplastic and apoplastic.

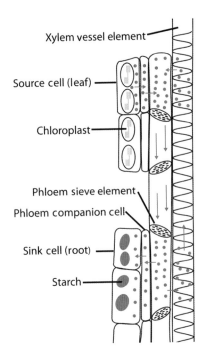

Figure 5.24 Translocation of sucrose from source to sink through the phloem is by pressure flow. As sucrose is loaded into the phloem at a source, water uptake occurs by osmosis and creates local high pressure. With the unloading of sucrose at the sink, water also leaves the phloem, lowering the pressure in the phloem. These differences in pressure cause sucrose to move from sources to sinks within the plant.

there is respiration but not photosynthesis, there is a need to import sucrose from the shoot of the plant to support growth and respiration.

Photosynthate that builds up as starch in the chloroplasts of leaf mesophyll cells during the day is made into sucrose and translocated to sinks or storage tissue at night. The active accumulation of sucrose in the phloem at the source (leaf) requires energy from ATP and attracts water by osmosis, which also enters phloem sieve tube cells and creates pressure which is higher than anywhere else in the phloem system (Figure 5.24). At the various "**sinks**" within the plant, such as roots that are respiring, or developing fruits or grain, the sucrose is unloaded from the phloem and used for growth or respiration. As sucrose is removed from the phloem, water will also leave by osmosis as the solute concentration decreases, and this becomes a region of low pressure in the phloem system. Therefore, flow occurs through the phloem from the source region toward areas where pressure is lower due to the use of sugars (Figure 5.24). Since flow through the phloem is from regions of high pressure to regions of low pressure, sucrose is continuously translocated from sources to sinks within the plant.

Chapter 6

Reproduction in flowering plants

The flowering plants—angiosperms—evolved more recently than the gymnosperms, which also produce seeds. The innovation for which angiosperms are known allows them to interact with pollinators, which in turn help angiosperms broaden their genetic variation. The fruits that develop from the ovaries of their flowers provide many means of dispersal to move plants into new environments. The variations in flower and fruit structure related to pollination and dispersal are discussed in this chapter.

Flower structure

A **complete flower** (Figure 6.1) has four parts: a calyx, a corolla, and staminate and pistillate flower parts attached at a **receptacle**. Flowers develop from apical meristems or axillary buds, and are sometimes compared to a short stem with several nodes and unelongated internodes to which its structures are attached.

The lower or outermost flower part is a whorl (circle) of **sepals** collectively called the **calyx**. The sepals are often green, but may be colored like petals. Their function is to protect the developing flower bud from physical injury, most importantly to prevent desiccation (drying out) of the tender tissues. A mature flower may already have lost its calyx, but in other cases, structures that appear to be petals may actually be sepals.

Above the calyx is the whorl of **petals** on the flower; the petals are collectively called the **corolla**. These are often large and colorful, and their arrangement can determine and help attract the type of pollinator that will be compatible with the flower of a particular species. There are often the same number of petals and sepals and they may appear to alternate when viewed from the center of the flower.

The number of petals and sepals often corresponds to the number of other flower parts, such as the number of stamens or chambers in the ovary. There

90 Structure and Function of Plants

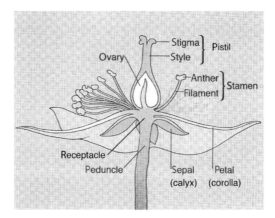

Figure 6.1 A diagram of a complete flower with a superior ovary. (Drawing from Wilkins 1988, used with permission.)

is even a collective term for the corolla and calyx together, the **perianth**. In some flowers, such as tulips, the flower "petals" are not differentiated into petals and sepals, and so the perianth is, in this case, made up of "tepals" (Figure 6.2).

Figure 6.2 Tulip with tepals which are hybrids of petals and sepals.

The staminate (male) and pistillate (female) flower parts contain the male (**pollen**) and female (**egg**) reproductive cells (or gametes) that can unite to create a new individual and develop into an embryo. The staminate flower parts (Figure 6.1) consist of the **anthers**, which usually have four chambers, and the **filament**, the stalk that gives an anther height and by which it is attached to the receptacle. An anther and filament together is referred to as a **stamen**.

The pistillate flower parts (Figure 6.1) consist of one or more pistils. Each **pistil** consists of one or more ovaries, which in turn contain one or more chambers. Located within each chamber of the ovary may be one or more ovules. Each ovule will house one egg and develop into one seed if fertilized. Each pistil also has a **stigma**, which is a sticky pollen receptacle, and a **style**, the stalk that provides height to the stigma. If a flower has more than one ovary attached to a single receptacle, the ovaries can develop together to produce an **aggregate fruit** like raspberry or blackberry. If several flowers with

Figure 6.3 Flower of fuchsia, which has an inferior ovary. With one of the petals removed, the style can be seen between the ovary, which is green, and the attachment of the petals and stamens.

single ovaries fuse together at maturity, they develop together to produce a **multiple fruit** such as pineapple.

The **ovary** may be attached above the other three main flower parts, in which case it is termed a superior ovary (Figure 6.1), or the other flower parts may be attached above the ovary, in which case the ovary is termed inferior (Figure 6.3). The location and number of flower parts vary tremendously, and may also be of an intermediate type.

An **incomplete flower** is one that is missing one of the four parts of a complete flower. A **perfect flower** is one that has both staminate and pistillate flower parts, but may be missing the corolla and/or calyx, and therefore be incomplete. Grass flowers (Figure 6.4) have the staminate and pistillate flower parts but no calyx or corolla, and are therefore perfect but incomplete.

An **imperfect flower** is one that does not have both staminate and pistillate parts, and so therefore could not be complete. An example of a plant that has separate male and female flowers is maize, in which the tassel (Figure 6.5a) is the male inflorescence, and each ear (Figure 6.5b) is a female inflorescence. If staminate and pistillate flowers are borne on the same plant, as in maize, the species is **monecious**; if they are borne on separate plants, the species is termed **dioecious**. Buffalo grass, a drought-tolerant warm-season grass native to the short-grass prairie of the southern Great Plains produces separate plants with either male (Figure 6.6a) or female (Figure 6.6b) flowers.

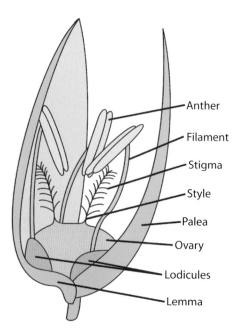

Figure 6.4 A grass flower, which has male and female flower parts but no sepals or petals, making it perfect but incomplete. The lodicules of grass flowers absorb water and expand to open the enclosing bracts, the lemma and palea. This allows the three stamens to fall out of the flower and shed pollen to the wind, while the two feathery stigmas are exposed to wind-carried pollen from other flowers of the same species.

(a) (b)

Figure 6.5 The tassel (a) is the male inflorescence of a maize plant, located at the top of the stem where it developed from the apical meristem (Figure 4.9). The ear (b) is the female inflorescence of maize, and developed from an axillary bud. The silks are the styles, and there is one for every kernel. Each fertile flower on the corn ear must be fertilized for a kernel to develop.

(a) (b)

Figure 6.6 The inflorescence of (a) a male buffalo grass (*Bouteloua dactyloides*) plant and (b) a female buffalo grass plant. Note that although these grasses are wind-pollinated, the anthers of the flowers on male plants and the stigmas of the flowers on female plants are brightly colored.

94 *Structure and Function of Plants*

Figure 6.7 While the showy red "petals" of poinsettias are actually leaves, the flowers at the stem apex are small and inconspicuous.

In many cases, flowers have altered these basic structures to attract pollinators. The flowers of poinsettias are yellow and inconspicuous (Figure 6.7), but the upper leaves have been turned into colorful bracts that act as large, red petals. And a native plant, columbine, has developed spurs or deep depressions in each of its five petals that contain a **nectary**, a gland that secretes nectar (Figure 6.8).

Inflorescences

Flowers are borne on a **peduncle** (Figure 6.1) either singly or in a group called an **inflorescence**. The primary variation in inflorescence structure is in the branching pattern (Figure 6.9). An inflorescence in which flowers are attached directly to the main axis of the inflorescence (or **rachis**) without a pedicel (stalk) is a **spike** (Figure 6.10a). When flowers are attached to the main axis by way of a pedicel, the inflorescence is a **raceme** (Figure 6.10b). When groups of flowers are borne on branches within the inflorescence, the

Figure 6.8 The petals (a) of columbine have deep spurs (b) with nectaries to attract pollinators.

inflorescence is termed a **panicle** (Figure 6.10c). The seed head of rice is a panicle. If the flowers are all borne on pedicels of the same length that are attached at a single point, resembling the underside of an umbrella, the inflorescence is called an **umbel** (Figure 6.10d). An inflorescence such as sunflower is termed a **composite head** (Figure 6.10e). Heads are composed of many fertile disk flowers that develop into seeds, surrounded by irregular, infertile ray flowers with large petals.

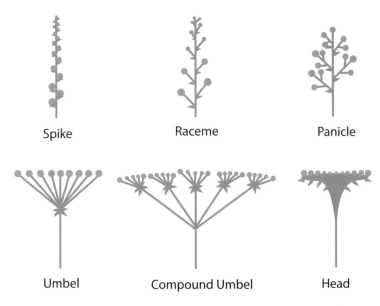

Figure 6.9 Diagrams representing the arrangement of individual flowers (pink disks) on different types of inflorescences.

Figure 6.10 Examples of the inflorescence types represented in Figure 6.9: (a) spike of crested wheatgrass (*Agropyron cristatum*), (b) raceme of sainfoin (*Onobrychis viciifolia*), (c) panicle of smooth bromegrass (*Bromus inermis*), (d) compound umbel of wild carrot or Queen Anne's lace (*Daucus carota*), and (e) the head of a sunflower (*Helianthus*) bisected to reveal the attachment of flowers to the receptacle, with fertile disk flowers filling the center of the head, and sterile ray flowers around the margin.

Vegetative reproduction

Reproduction may occur through the growth of axillary buds on stolons or rhizomes (Figures 4.4a and 4.4b), and other vegetative parts may produce new plants. These are examples of vegetative reproduction, which results in a new plant that is genetically identical to the original, and is termed a **vegetative clone**. Most offspring of dandelions (*Taraxacum*) are clones of the parent plant, because fertilization is not required for seed production in this species. Quaking aspen (*Populus tremuloides*) reproduces vegetatively, with new trunks arising from the root system by suckers.

The largest organism in the world may be a clonal colony of quaking aspen with over 60,000 trunks in the Fishlake National Forest of the Wasatch Mountains of Utah. It covers over 100 acres and is named Pando, Latin for "I spread." When fires destroy the top growth of the plant, the root system sends up new stems. While individual trunks may be more than 100 years old, the plant itself is thought to be at least 80,000 years old.

Pollination

For pollination to occur, pollen must be carried from the anther to the stigma. Pollen movement can occur by wind, water, insects, birds, or mammals. Grasses and gymnosperms use wind pollination. These plants typically do not have showy petals or nectaries, which are expensive to produce in terms of photosynthate. The trade-off is that these plants must produce enormous quantities of lightweight pollen (e.g., 50,000 pollen grains per flower; Figure 6.11). The stigmas of grass flowers are relatively large, and their stamens

Figure 6.11 A cloud of pollen from the flowers of a male buffalo grass plant.

Figure 6.12 A bee (yellow circle) deeply involved in the pollination of a flower of a foxglove inflorescence. (Photograph by V. Clarke, used with permission.)

are well exposed (Figure 6.4). Each grass flower has only one ovary with a single ovule, but many flowers are packed onto a single grass inflorescence (Figure 6.10c)

Insects are the most common animal pollinators, and bees are the most common insect pollinators (Figure 6.12). Bees are most often attracted to red flower color, and gather both nectar and pollen. Some flowers have strong scents or visible nectar guides (Figure 6.13) to attract birds or insects. Other flowers use nectar guides that resemble a target created by UV reflectance, which can be seen by bees, with darker shades toward the center of the flower like a bull's-eye (Figure 6.18).

Flowers use many mechanisms to assure **cross-pollination**, or the movement of pollen from one flower to another to avoid inbreeding and increase the variability of their offspring. In the case of alfalfa (*Medicago sativa*), the stamens are fused together around the pistil and enclosed in two petals called the keel (Figure 6.14a). When a bee lands on the flower to collect nectar, the keel ruptures and the staminate column springs up (Figure 6.14b), smacking

Figure 6.13 Penstemon flowers with pollen guides and a flaring lower corolla to act as a landing platform for pollinators.

(a) (b)

Figure 6.14 An alfalfa inflorescence with untripped flowers (a) and with one tripped flower (b). The stamens and pistil are enclosed in the keel, a pair of fused lower petals. The weight of the bee landing on the keel causes it to rupture and the staminal column springs up against the bee. It can be seen upright against the large standard petal.

the bee on the abdomen and covering it with pollen. At the same time, the stigma is dusted with the collection of pollen from all the other flowers that the bee has visited, so each ovule may be fertilized by pollen from a different flower.

Some mammals, such as bats, act as pollinators of light colored, strongly fruity scented flowers that open at night, and birds, especially hummingbirds, pollinate the flowers from which they collect nectar. These flowers often are red or yellow, and have long, fused corolla tubes with nectar deep inside, where only the tongue of a hummingbird can reach.

Fertilization

Pollen grains contain two cells, a **tube cell** that forms the pollen tube, and a **generative cell** that divides and forms two **sperm cells** (Figure 6.15a). Pollination occurs when a pollen grain from an anther is transferred to the stigma of a receptive pistil of the same species (Figure 6.15b). If the pollen and the stigma are compatible, the pollen grain will germinate on the stigma and the tube cell will form a pollen tube. The wall of a pollen grain contains pores or weak spots through which the **pollen tube** emerges if the pollen grain germinates on the stigma. The generative pollen cell divides to form two sperm cells during migration down the pollen tube. When the pollen tube reaches the ovule, it enters through a small opening called the **micropyle**, which is often still visible on a seed.

In the ovule, an embryo sac develops, and a number of nuclei are formed by division of the original single cell (Figure 6.15c); one of these is the egg. **Fertilization** (Figure 6.15d) occurs when one of the two sperm cells unites with the egg to form the **zygote**; the zygote will develop into the embryo. The second sperm cell fuses with two (female) polar nuclei to form the endosperm, which is nutritive tissue used by the developing embryo. The entire ovule develops into the seed. If there is more than one ovule within the ovary, each must be fertilized independently of the others. Examples of a single ovary with multiple ovules that develop into seeds are the pods of legumes, such as beans (*Phaseolus*) and peas (*Pisum*; Figure 6.16). The pod is the ovary, and each of the seeds inside has developed from a separate ovule.

Dispersal of seeds

When seeds mature, they may simply fall to the ground and develop into new plants, but this would create considerable competition for the mother plant and would slow the rate of adaptation and eventual development of new species by limiting the range of a given species. Therefore, plants use numerous devices to disperse their seed.

One dispersal mechanism that humans have exploited for their own use is the breeding of the ovary into edible fruit. When we consume the flesh

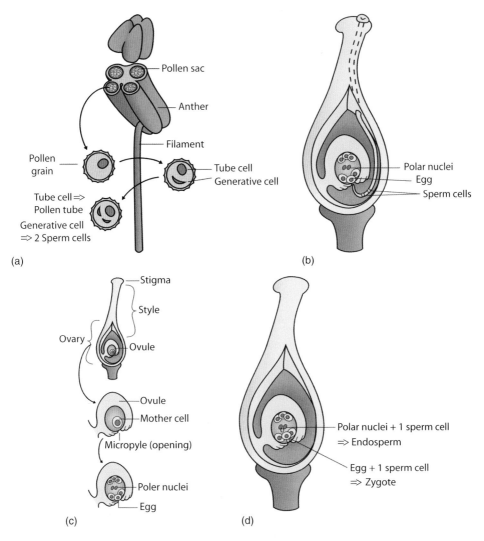

Figure 6.15 (a) Anthers usually have four chambers filled with pollen grains. Each pollen grain initially contains one cell with a single set of chromosomes. This cell divides once to produce a tube cell, which will form the pollen tube, and a generative cell. The generative cell divides again before fertilization. (b) Pollination occurs when a pollen grain is transferred to the stigma of a compatible flower. Germination of the pollen grain is followed by growth of a pollen tube through the style to the ovule. (c) In the immature seed (ovule), several cells are formed that have only one set of chromosomes. One of these is the egg and others (the polar nuclei) function to produce nourishment for the developing embryo. (d) When the pollen tube reaches the ovule, one of the sperm cells formed from the generative cell fuses with the egg, forming the zygote, and the other fuses with the polar nuclei to form the endosperm. The zygote develops into an embryo, and the ovule matures into a seed.

of cherries, peppers, oranges, and watermelons, the seeds are discarded but often at a considerable distance from the source of the fruit. Nuts and berries (Figure 6.17a) are also dispersed by animals that discard the seeds or bury them for later use. Some buried seeds will not be utilized and will germinate where they were cached.

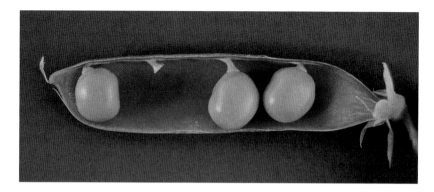

Figure 6.16 Peas (*Pisum*) are legumes in which several ovules (seeds) develop in a pod (ovary). One pea has been removed to show the funiculus, the stalk to which it was attached during development.

Figure 6.17 Some examples of dispersal mechanisms: (a) hawthorn (*Crataegus*) berries are eaten by birds, (b) achenes of salsify (*Tragopogon*) are light and can travel long distances on the wind, and (c) samara of maple (*Acer*) are dispersed short distances by wind. The pods of some legumes (d), such as this birdsfoot trefoil (*Lotus corniculatus*), disperse seeds by splitting open forcefully along their seams as they become dry. Note that the calyx of the flowers can still be seen at the base of the elongating pod, while the petals are still attached to the tip of the elongating pod.

Some plants use wind to disperse their seed which may have plumes that stay aloft on the wind (Figure 6.17b), or have wings that carry the seeds away from the mother plant (Figure 6.17c).

Legume fruits consist of a pod that splits in half along its margins due to tension that develops as the pod dries (Figure 6.17d). The two halves of some of these pods are under so much tension as they dry that seeds are dispersed with explosive force.

Water is used for seed dispersal by some plants, with palm as an example of a seed that can be carried enormous distances on ocean currents. Even in dry environments, such as an arid arroyo, the occasional rush of floodwaters can be used to disperse seeds a long distance from the mother plant.

Seed dispersion in a rare native species

Flannel bush (*Fremontodendron*) shrubs are evergreen chaparral species native to California, Arizona, and Baja California. One of the three species of flannel bush, *F. decumbens*, is rare—found in only one county in California—and its seed distribution has been studied to aid in its conservation (Boyd and Serafini 1992; Boyd 2001).

Pollination in *F. decumbens* is promoted by fluorescence of both the pollen and nectar (Figure 6.18; Boyd 1994); however, flower and seed production in *F. decumbens* is reduced by insect or rodent predation at every stage of its development. While the production of flower buds on the plant is abundant, about 80% of the buds are killed by a moth larva that bores a hole in the sepals and feeds on the bud. Twenty percent of the flower buds survive, but one of every two of the resulting flowers is killed by larvae of the same moth that attacks the buds. The flowers that survive produce fruit, but fruit mortality is about 80%. The larva of a different moth bores into the fruit and consumes the developing seeds. With all this predation, only about 100 seeds per plant are produced in a given year.

The seeds of this shrub are relatively large and therefore desirable as food. Of the seeds that fall to the ground, rodents consume about 90% and of the seedlings that do germinate, 60% are eaten by rodents or insects. The one bit of help the seeds of this plant species receive comes from native harvester ants. While these ants are also interested in the seeds for food, they consume the elaiosome (Figure 6.19), an appendage on the outside of the seed that is rich in lipids. The ants carry flannel bush seeds to their nests, consume the elaiosome, then discard the undamaged seeds either in their nests or outside on middens (trash heaps) near the nest's entrance. Flannel bush invests nearly 10% of its photosynthate in the creation of the elaiosome (Boyd 1996) and it must therefore receive some benefit. Fire (or heat treatment followed by chilling) is required for seed germination, and removal of the elaiosome by harvester ants does not harm the seed—the pericarp is three times thicker at the elaiosome—but does not improve its germination, either. The apparent benefits are indirect. One is the transportation of the seeds away from the shrub canopy, where they are more easily found by rodents, and where they would have to compete with root sprouts from the mother plant following fire. Harvester ant nests are in open areas between shrubs, and while the environmental conditions on middens are not particularly favorable, seeds remaining in the ants' nest are protected from rodent predation. If they do germinate, the developing plant should benefit from a nutrient-enriched soil environment. The second benefit of ants removing the elaiosome is that while rodents are after the seed itself and will discard the elaiosome, the elaiosome appears to produce volatiles that attract rodents. After its removal, even if the seeds are in the open, they are harder for rodents to find.

104   Structure and Function of Plants

Figure 6.18 Flowers of flannel bush (*Fremontodendron*) photographed in visible light without a filter (above) and with a filter showing absorbance of light in the UV range (300–400 nm); a gray scale is included with both photographs. Solitary bees that pollinate these flowers can sense UV light and are guided to the style, stamens, and nectaries as dark central structures surrounded by a bright ring of sepals. (Photograph from Boyd 1994, used with permission.)

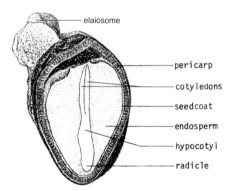

Figure 6.19 The elaiosome of this flannel bush (*Fremontodendron*) seed is consumed by ants without harming the seed. By carrying the seed away from the canopy of the mother

plant and consuming the elaiosome, seeds are not as likely to be consumed by rodents and therefore stand a better chance of survival. (Drawing from the USDA, NRCS 2007.)

Because in nature the seeds of these species require fire to germinate, fire suppression also suppresses germination, sexual reproduction, and therefore genetic variation. While insect and rodent predation are big challenges for seed production of this rare species, human residential encroachment and the resulting suppression of fire also inhibits the ability of the species to adapt to a changing environment through sexual reproduction.

Chapter 7
Plant nutrition

The root system of most plants absorbs water and mineral nutrients for the entire organism, just as the shoot carries out photosynthesis to support the entire plant. To understand plant nutrition, we need to know how the soil environment provides nutrients to plants, how the plant selectively takes nutrients from the soil, and the roles nutrients play once they are taken up by the plant.

Soil components

Soils are composed of mineral particles and **humus**, which is nonliving organic matter. Soils also contain differing proportions of air, water, and living organisms like earthworms.

Soil mineral particles are the products of weathering of the underlying rocks by natural forces such as glaciers, wind, and rain, and usually contain two or more elements, primarily aluminum, silicon, and magnesium.

Soil mineral particles

Soil mineral particles are classified by size from largest to smallest as sand, silt, and clay (Figure 7.1). **Sand** particles range from 2 to 0.02 mm in diameter, and soils dominated by sand have large pore spaces for air and water movement but little surface area relative to the volume of particles. Therefore, sandy soils have good aeration and poor **water-holding capacity**. **Clay** particles, the smallest, average less than 0.002 mm in diameter, and clay soils have small pore spaces but relatively high surface area so they have very good water-holding capacity but poor aeration. **Silt** particles are intermediate in size, ranging in average diameter from 0.02 to 0.002 mm. Soils contain various mixtures of sand, silt, clay, and organic matter.

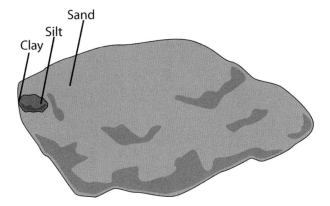

Figure 7.1 Relative sizes of soil mineral particles.

A thin film of water adsorbs (adheres) to the surface of mineral particles, so the water-holding capacity of a soil is proportional to the combined surface area of the particles of which it is composed. The smaller the particles, the greater the total surface area in a given volume of the soil, which is why a soil dominated by clay has a higher water-holding capacity than a soil dominated by sand. At the same time, the larger the mineral particles, the larger the air spaces between neighboring particles, and the better the aeration of the soil. Aeration is important for plant growth because plant roots require oxygen for respiration, and the oxygen must be acquired from air in the soil. The small particles of a clay soil pack together tightly, so the soil may contain a relatively high proportion of water, but the rate at which water can infiltrate into or drain from a clay soil is low. This can cause a higher rate of surface runoff of rain or irrigation water than would occur on a sandy soil. Dense clay soils also have relatively poor root penetration.

Soil organic matter

The humus or nonliving organic matter content of a soil will increase both the soil's water-holding capacity and aeration. Humus provides a natural glue that causes mineral particles to aggregate (Figure 7.2), providing structure to soils like that seen in the castings of earthworms. Humus is lightweight and spongy in nature, and will swell as it absorbs water then shrink as it dries. Humus also greatly improves the ability of a soil to provide nutrients to plants. Mineral soils range from 1 to 10% humus, while organic soils, which develop from swamps and marshes, contain more than 30% humus.

The organic matter of humus is mineralized (broken down to its component compounds) over time by soil microbes, releasing the nitrogen, phosphorus, and sulfur in organic molecules such as DNA and proteins, but also acidifying the soil. Because humus is constantly being consumed by microbes in mineral soils, it must be replaced through such practices as adding compost to a garden soil, or plowing down crop residues. In agricultural soils

	Before wetting		After wetting	
	High OM	Low OM	High OM	Low OM

Figure 7.2 On the left of the figure, dry soils with high or low organic matter (OM) appear to have similar structures. However, when both soils are wetted (on the right side of the figure), the structure of the low OM soil is lost, while the structure of the high OM soil is retained. (Figure from Brady 1990, page 113. Reprinted by permission of Pearson Education, Inc., Upper Saddle River, NJ.)

used to grow annual crops, cultivation increases aeration and turning over the surface soil increases the soil temperature, which increases the activity of microbes and the rate of organic matter mineralization. In uncultivated ecosystems, leaf fall and animal droppings slowly replenish the organic matter content of soils. In organic farming systems, incorporating livestock manure (Table 7.1) or a living mulch high in protein such as an annual clover or vetch, provides both nitrogen and organic matter but mineralization of the nitrogen is dependent on the activity of soil microorganisms. The perennial sods of the Great Plains have high natural fertility because of the extensive grassroot systems that developed and then were mineralized over millennia.

The best soils for plant growth are those classified as loams, having about equal proportions of sand, silt, and clay. Their water-holding capacity is

Table 7.1 Nutrient content of manure. Data from Thorup 1984. The nutrient content of most manures is very low as a percentage of dry weight compared to chemical fertilizers, but manures include valuable organic matter. Note that the nitrogen (N) content of poultry mature is proportionally high and that the phosphorus (P_2O_5) content of feedlot beef cattle manure is higher than that of dairy cattle manure. This is because beef cattle in feedlots are fed more grain while dairy cattle are fed more hay, and grain is higher in phosphorus than forages.

Manure source	N (%)	P_2O_5 (%)	K_2O (%)
Beef (feedlot)	0.71	0.64	0.89
Dairy cattle	0.56	0.23	0.60
Hog	0.50	0.32	0.46
Horse	0.69	0.23	0.72
Sheep	1.40	0.48	1.20
Chicken (no litter)	1.56	0.92	0.42

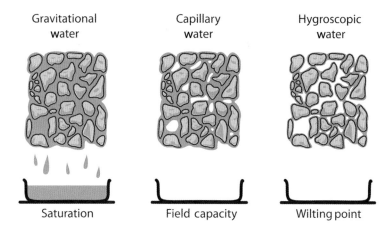

Figure 7.3 Gravitational water will drain from a soil saturated with water for a day or more, leaving macropores filled with air. In a soil at field capacity, the smaller micropores are filled with capillary water that can be used by plants. Even when a soil has reached the permanent wilting point, where plants have withdrawn all the water they can, a soil can still retain significant hygroscopic water adsorbed to soil particles. (Figure from Brady 1990, page 143. Reprinted by permission of Pearson Education, Inc., Upper Saddle River, NJ.)

good, but aeration and root penetration is also good on these soils. Plants can be grown in containers in rooting media designed to be light and well drained while still having good nutrient- and water-holding capacities. Components of these rooting mixtures include sphagnum peat moss, vermiculite, and sand.

Soil water

The water in soils is classified as gravitational water, capillary water, and hygroscopic water (Figure 7.3). **Gravitational water** is the excess water that will drain from a soil due to gravity. Nutrients are leached by gravitational water, so it is not of much value to plants. The water that remains following drainage is called **capillary water**. This water is loosely held in small pores by surface tension, and is easily used by plants. Most nutrients used by plants are dissolved in capillary water. The water that is not available to plants because it adheres so tightly to the surface of mineral particles is called **hygroscopic water**. The hygroscopic water content of clay soils is commonly 15% and that of a sandy soil is as little as 3%.

Acidity and alkalinity

Water molecules can break apart into H^+ and OH^- ions (protons and hydroxyl ions, respectively). If a solution contains more H^+ than OH^-, it will be acidic; if there is more OH^- than H^+, it will be alkaline. If a strong acid like hydrochloric acid (HCl) is added to a solution, it can separate into its

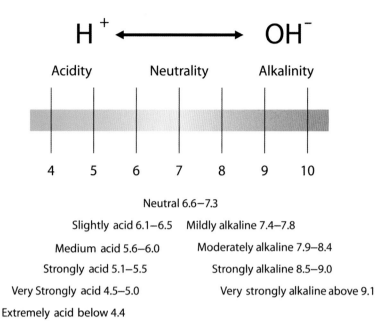

Figure 7.4 The pH scale is a measure of acidity or alkalinity of aqueous solutions, including the soil solution. Characterization of soil pH from Martin et al. (1976).

components, H^+ and Cl^-, acidifying the solution. If a base like sodium hydroxide (NaOH) is added to a solution, it will separate into Na^+ and OH^-, and make the solution more alkaline by increasing the number of OH^-. The **pH** scale (Figure 7.4) is an expression of the number of H^+ in solution, and like the Richter scale that measures the strength of an earthquake, a change of one unit on the pH scale is equivalent to a 10-fold change in H^+ concentration.

A pH of 7 is chemically neutral, and acidity increases as pH decreases from 7 to 1, while alkalinity increases as pH increases from 7 to 14 (Figure 7.4). Some plants that require acidic soils (pH 4.5–5.5) are blueberry, azalea, ferns, rhododendron, and camellia. Plants that require alkaline soils (pH 7.5 and higher) include alfalfa, asparagus, spinach, beets, and lettuce. Hydrangea flower color reflects the pH of the soil—blue in acidic soils, and pink in alkaline soils.

Soils in higher rainfall regions such as the eastern United States tend to be acidic because nutrient cations are leached from the soil, leaving more H^+ (low pH), while soils in low-rainfall regions tend to retain an excess of nutrient cations and relatively little H^+ in solution, resulting in an alkaline (high) pH. The pH of a soil can be made more acidic through the application of sulfur, although this impractical on a large scale, but acidic soils used for cropping in the eastern United States are routinely made more alkaline through liming.

Liming is the application of calcium carbonate (ground limestone) to acidic soils, which have an excess of H^+. When the calcium carbonate ($CaCO_3$)

mixes with water (H_2O) in the soil, the products are Ca^{2+}, HCO_3^-, and OH^-. The OH^- released from the limestone combines with excess H^+ from the soil solution producing water and thereby reducing the concentration of H^+, raising the pH of the soil.

Soil nutrients

The nutrients in soils are ions, which are charged atoms or molecules that can exist as salts like sodium chloride (NaCl), but which separate when dissolved in soil water (Na^+ and Cl^-). Many of the nutrients required by plants are positively charged ions, or **cations** (ammonium as NH_4^+; potassium as K^+; calcium as Ca^{2+}; magnesium as Mg^{2+}). Other important nutrients are negatively charged ions or **anions** (nitrate as NO_3^-; phosphorus as $H_2PO_4^-$ or HPO_4^{2-}; sulfate as SO_4^{2-}). Nitrate is mobile in the soil solution and can

Table 7.2 Nutrients required by plants. The amount of nutrients required by plants, from highest to lowest, relative to molybdenum. The actual concentrations by dry weight of plant nutrients are based on analyses of the shoots of many plant species, and range from 6% for hydrogen to 0.1 ppm for molybdenum.

	Atoms of element relative to molybdenum[a]	Form of uptake[b]	Mobility in the phloem[c]
Macronutrients			
Hydrogen	60,000,000	H_2O	N/A
Carbon	40,000,000	CO_2	N/A
Oxygen	30,000,000	CO_2, O_2	N/A
Nitrogen	1,000,000	NO_3^-, NH_4^+	Mobile
Potassium	250,000	K^+	Mobile
Calcium	125,000	Ca^{2+}	Immobile
Magnesium	80,000	Mg^{2+}	Mobile
Phosphorus	60,000	$H_2PO_4^-$, HPO_4^{2-}	Mobile
Sulfur	30,000	SO_4^{2-}	Immobile
Silicon[d]	30,000	$Si(OH)_4^0$	Immobile
Micronutrients			
Chlorine	3,000	Cl^-	Mobile
Iron	2,000	Fe^{2+}, Fe^{3+}	Immobile
Boron	2,000	$B(HO)_3^0$	Immobile
Manganese	1,000	Mn^{2+}	Immobile
Sodium[d]	400	Na^+	Mobile
Zinc	300	Zn^{2+}, $Zn(OH)_2^0$	Mobile
Copper	100	Cu^{2+}	Immobile
Cobalt[e]	2	Co^{2+}	N/A
Nickel	1	Ni^{2+}	Immobile
Molybdenum	1	MoO_4^{2-}	Mobile

[a] Epstein and Bloom (2005).
[b] Bennett (1993).
[c] Taiz and Zeiger (2006).
[d] Not essential for all plants (Epstein and Bloom 2005).
[e] Required for nitrogen fixation (Epstein and Bloom 2005).

be leached from a soil in gravitational water by heavy precipitation or irrigation. Phosphates become bound to elements in soil mineral particles and therefore leach less readily than NO_3^-, but can be lost in runoff from heavily fertilized or manured soils (Table 7.2).

Both the clay particles and humus of soils are negatively charged, and form relatively loose ionic bonds between nutrient cations (+) and the soil (−), holding cations against leaching while still making them available for plants (Figure 7.5). The ability of a soil to provide nutrients to plants can be characterized as the soil's **cation exchange capacity** or CEC. Sand does not carry a significant negative charge, and therefore sandy soils have a low nutrient-holding capacity or CEC.

Nutrients can be removed from the soil by uptake into the plant root. This can occur where roots or root hairs contact the soil, or nutrient ions may move with soil water into and through root cell walls as far as the endodermis (Figure 7.6). The walls of cells of the endodermis are imbedded with suberin,

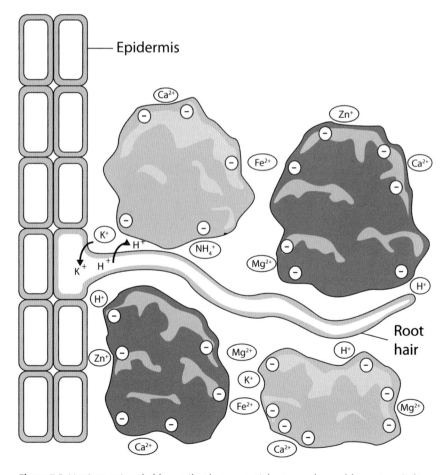

Figure 7.5 Nutrient cations held on soil or humus particles are exchanged for protons (H^+) from the plant root. A similar exchange of OH^- for nutrient anions occurs, allowing the plant to maintain a constant internal pH.

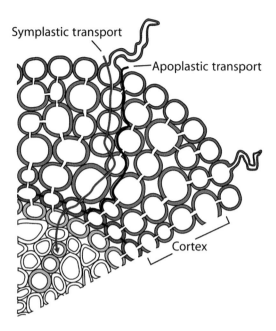

Figure 7.6 The apoplastic route of nutrient uptake includes the cell wall and intercellular spaces of the epidermis, the cortex, and the endodermis. Nutrient uptake into the cytoplasm can occur in any of these cells, which lie outside the stele, after which nutrient transport proceeds symplastically through the cytoplasm and plasmodesmata of successive cells. To cross into the vascular tissue inside the stele, nutrient ions must move through the symplast by way of plasmodesmata from the endodermis to xylem parenchyma cells, which then load nutrients into nonliving xylem elements.

which makes them impenetrable to water and therefore to nutrient movement. Uptake from the continuous cell wall space (apoplast) into the system of connected cellular protoplasms (symplast) is through protein channels in the plasma membrane of root cells in contact with the soil solution.

Macronutrients

Macronutrients are those nutrients that are used in large quantities by plants, constituting more than 0.5% by weight of plant dry weight. The most abundant elements in plants are carbon, hydrogen, and oxygen, which are also the major elements in all organic (carbon-based) molecules. Hydrogen is acquired from water (H_2O), causing plants to evolve O_2 gas as a by-product of photosynthesis. Carbon and oxygen, which come from CO_2, are incorporated into plants in the process of photosynthesis.

The other elements essential to plant growth can be acquired by the plant from the soil solution. See Table 7.3 for the pounds of nutrients removed in 1 year from 1 acre (0.4 ha) of land by an alfalfa crop. The nutrients most commonly added to soil or rooting media to support plant growth are nitrogen (N), phosphorus (P), and potassium (K). The natural abundance of

Table 7.3 Mineral nutrients removed in crops. Data from Thorup 1984. The quantity of mineral nutrients removed by the harvest of a 10-ton crop of alfalfa (and the quantity of nitrogen fixed symbiotically) produced on 1 acre (0.4 ha) of cropland in 1 year. The mineral nutrients have been mined from the soil by this crop, which is harvested several times each season and may be sold off the farm. Depending on the soil's nutrient reservoirs, some of these elements will have to be replaced using chemical or organic fertilizers if this cropland is to continue to be productive.

Nutrient	lbs (kg) per 10 tons alfalfa
Nitrogen	530 (240)
Phosphorus (as P_2O_5)	75 (34)
Potassium (as K_2O)	550 (249)
Calcium	320 (145)
Magnesium	66 (30)
Sulfur	48 (22)
Zinc	0.3 (0.14)
Iron	4.0 (1.81)
Manganese	2.0 (0.91)
Copper	0.2 (0.09)
Boron	0.6 (0.27)
Molybdenum	0.01 (0.0045)
Chlorine	3.0 (1.36)

these nutrients in soils of the continental United States is shown in Figure 7.7. Nitrogen is naturally high in the cold, wet soils of the north central region, where the formation of organic matter is abundant but its mineralization is slow. In the warm, dry soils of the semiarid western US, organic matter and therefore N in soils is naturally low, but P and K are naturally high.

Nitrogen

Nitrogen (N) is an essential component of all amino acids, the building blocks of proteins. All enzymes are proteins (Chapter 9), and therefore N is required for all cellular metabolism (e.g., photosynthesis and respiration—Chapters 10 and 11). Proteins are also responsible for the selectivity of membrane transport because they comprise the channels through which elements move in and out of cells. Nitrogen is a key component of chlorophyll, the pigment that imparts the green color to plants, and which is essential for the transfer of energy from sunlight into carbohydrates through photosynthesis. The four bases (adenine, guanine, cytosine, and thymine) that constitute the genetic code found in DNA are nitrogenous compounds. It is N that gives these compounds their basic (positively charged) nature.

N is **mobile** in the plant, meaning it can be transported in the phloem, and therefore the most pronounced early symptom of N deficiency is yellowing or **chlorosis** of *older* leaves. Because N is critical for new growth, the plant breaks down proteins in older leaves and sends N to younger leaves when it is in short supply. Mobile nutrients will also be salvaged from the leaves of healthy perennial plants before abscission (leaf drop) in autumn.

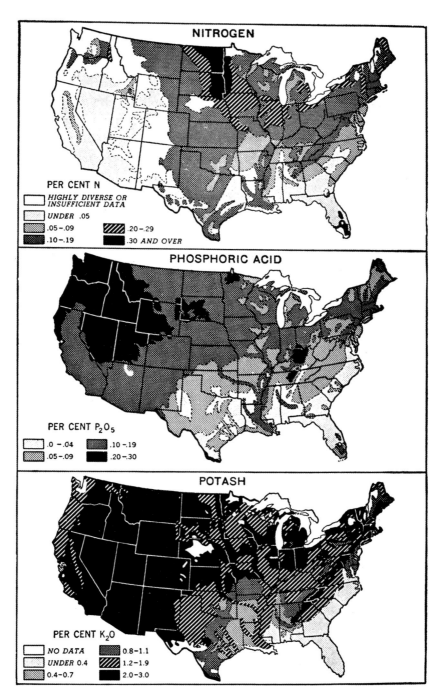

Figure 7.7 USDA maps of nitrogen (N), phosphorus as P_2O_5 (middle), and potassium as K_2O (bottom) in the surface foot of soil. Shading from lightest to darkest represents concentrations from 0 to 0.3% for N and P_2O_5, and from 0 to 3% for K_2O.

Figure 7.8 Application of nitrogen was heavier in the two horizontal stripes that can be seen in this Kentucky bluegrass (*Poa pratensis*) lawn. The grass responded to higher availability of N with dark green color and lush growth.

The yield of crop plants can be dictated by the amount of N that is applied if water and other nutrients are not limiting, because vegetative growth is roughly proportional to N availability (Figure 7.8). Nitrogen is taken up by plants as NO_3^- (nitrate), which is readily leached from the soil, or as NH_4^+ (ammonium). While plants can utilize NH_4^+ without the energy-demanding biochemical steps required for the use of NO_3^-, too much NH_4^+ at one time can be toxic. Furthermore, NH_4^+ is usually not available for long even when NH_4^+-containing fertilizers are applied to the soil, because NH_4^+ is quickly converted to NO_3^- by soil microbes. Ammonium is the higher-energy or "reduced" form of soil N, and NO_3^- is the lower-energy or "oxidized" form. Soil microbes can utilize the energy released when they convert NH_4^+ to NO_3^-. Nitrogen can also be acquired by legumes and some other plants from the air, which contains about 78% N as the gas N_2, through associations with soil-living bacteria (Chapter 3).

Potassium

Potassium (K) is used in large quantities in plant cells as a compatible solute—one that can occur in the cytosol in relatively high concentrations without a detrimental effect. Almost all other ions are excluded from the cytosol by sequestration in the vacuole, the cell wall, or other cell compartments. The function of guard cells depends on K^+, accompanied by both inorganic and organic anions (Talbott and Zeiger 1998). Potassium is also used as a counterion of nucleic acids, and is an important activator of proteins; the content of potassium and protein in plants are well correlated (Blevins 1984). With

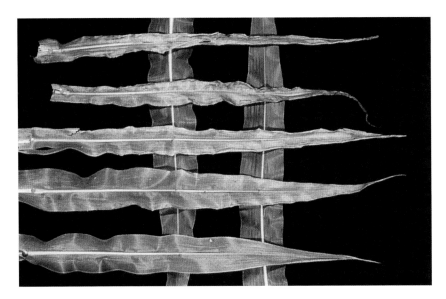

Figure 7.9 Maize leaves with the potassium deficiency symptom of marginal chlorosis and, on more mature leaves, marginal necrosis. (Courtesy of the Western Plant Health Association.)

K deficiency, plants such as maize may exhibit marginal leaf chlorosis (Figure 7.9). Potassium is also needed for the storage and utilization of carbohydrates and proteins in the roots of perennial plants. In alfalfa, a classic symptom of K deficiency is white spots on leaflet margins, but K^+ deficiencies in this species are also seen as a general reduction in winter-hardiness (Smith *et al.* 1986). Potassium-deficient perennial plants may not, in fact, be killed by low winter temperatures, but suffer from diminished spring regrowth caused by insufficient stored carbohydrates (Dhont *et al.* 2002) and proteins (Dhont *et al.* 2003).

Calcium

Calcium (Ca) in the cytosol is a potent activator of cellular responses. Therefore, cellular Ca is sequestered in the vacuole, the endoplasmic reticulum, and other organelles, or localized on the outside of the plasma membrane where it is required for selective ion transport. A relatively high concentration of Ca is found in the middle lamella, where it cross-links pectin molecules. Plant Ca concentrations are relatively high in meristems and young tissues. Therefore, a Ca deficiency is seen as damage to growing points in roots and shoots, and blossom-end rot of fruits.

Magnesium

Magnesium (Mg) is a component of chlorophyll, so a Mg deficiency causes the synthesis of chlorophyll to be inhibited. A Mg deficiency appears as an interveinal chlorosis where only the veins of leaves still retain green color. An interveinal chlorosis may occur simply because symptoms appear gradually and cells near the veins have the best access to water and nutrients from the xylem. Magnesium also functions as an activator of enzymes in energy transfer, as part of ATP, in photosynthesis and respiration, and the synthesis of RNA and DNA. Magnesium is a mobile nutrient, so symptoms will appear first on older leaves.

Phosphorus

Phosphorus (P) as inorganic phosphate (PO_4^{3-}) is a component of RNA and DNA, and of ATP, the molecule used to transfer energy in enzyme reactions in the plant. Phospholipids, which contain a phosphate group, are the fundamental component of the lipid bilayer of cellular membranes. Sugar phosphates constitute intermediates in critical biochemical pathways such as photosynthesis and respiration. A deficiency of P is especially detrimental to young plants because it inhibits their initial rapid growth, and P is therefore often banded below the seeds of crops at planting. While P and K are both naturally high in lower rainfall soils, P is never very soluble in the soil solution. In alkaline soils, P forms insoluble compounds with Ca and in acidic soils with iron or aluminum; even at a relatively neutral soil pH of 6–7, it is only sparingly soluble. The most pronounced symptom of P deficiency is dark green or blue-green leaves, stunted growth of young plants, and on warm-season grasses like maize, purple coloration (Figure 7.10). **Mycorrhizae** are nonpathogenic associations between plant roots and soil fungi that are very common in nature and that involve invasion of the root by the fungal partner to obtain organic nutrients. Formation of mycorrhizae increases by many times the distance into the soil a root can explore for nutrients, and can thereby increase the rate of uptake of soil P by more than four times (Epstein and Bloom 2005).

Sulfur

Sulfur (S) is a component of cysteine and methionine, 2 of the 20 essential amino acids. These 2 amino acids are widely used in proteins because they impart special structural and metabolic properties. Sulfur is a component of several enzymes involved in the electron transport steps of photosynthesis and respiration, and is a constituent of coenzyme A, which is involved in an early step of respiration (Chapter 11).

In higher rainfall areas, S is released from soil organic matter; in dryer areas, S occurs in the soil as gypsum ($CaSO_4$). Sulfur deficiencies were not common in the past, but atmospheric sulfur dioxide from the burning of high-sulfur coal has been reduced, and the shift to fertilizer with higher purities of N, P, and K has reduced the S-content of commonly used fertilizers.

A sulfur deficiency causes chlorosis of leaves similar to a N deficiency. However, S is an immobile nutrient, so chlorosis will initially appear in the

Figure 7.10 Maize plant showing phosphorus deficiency symptoms of stunted growth and purple coloration. Symptoms are typically seen in spring, when young plants with small root systems are growing in cool soils. (Courtesy of the Western Plant Health Association.)

youngest leaves. **Immobile nutrients** such as S and iron must be available to plant roots throughout growth because they are not readily transported from mature older leaves to younger deficient tissues.

Micronutrients

Micronutrients are those nutrients used in small quantities by plants, constituting less than 0.5% by weight of plant dry weight.

Silicon

Silicon (Si) is required primarily by plants in the ancient Equisetaceae family (Taiz and Zeiger 2006) but is present in large amounts in the soil and is accumulated in the tissue of many plants. In crops such as sugarcane, rice, and wheat, Si aids in production by providing resistance to lodging, and in rice and some dicots, Si contributes to resistance to disease (Epstein and Bloom 2005).

Chlorine

Chlorine (Cl) is required for the earliest step of photosynthesis (Chapter 10), where water is split to supply electrons to the photochemical (light-driven) reactions, producing oxygen. The chloride ion (Cl^-) is also used as a counterion for K^+ in stomatal function. Chloride is readily available in the environment, so deficiencies are rare, but early symptoms include wilting of leaf tips and leaf discoloration (bronzing).

Iron

Iron (Fe), an immobile nutrient, is used in the synthesis of chlorophyll, and is needed for the conversion of NO_3^- and atmospheric N_2 to forms that can be used by plants. An early symptom of Fe deficiency is interveinal chlorosis of younger leaves (Figure 7.11); eventually, Fe-deficient leaves become entirely chlorotic. Iron-containing enzymes, in which Fe^{3+} is reduced to Fe^{2+} and then oxidized back to Fe^{3+} to carry out oxidation–reduction reactions, are required for critical electron transport steps in both photosynthesis and respiration. Iron becomes insoluble in high pH (alkaline) soils, and plant roots release H^+, **organic acids**, and other compounds to increase the, availability of this important nutrient. Fertilizer Fe is best provided to plants in a **chelated** form, bound to organic molecules that release the Fe cation at the root surface. The nutrients Fe, copper, and manganese may be effectively provided to plants as foliar sprays.

Figure 7.11 Leaf of aspen from a tree growing on iron-deficient soil, showing interveinal chlorosis.

Boron

Boron (B) functions in cell wall structure, cross-linking large structural carbohydrates. A general symptom of B deficiency is dry, brittle tissue in both shoots and roots.

Manganese

Manganese (Mn) is required (along with Cl) in the production of oxygen early in photosynthesis. It is also an enzyme activator, used in the Krebs cycle of respiration (Chapter 11). A symptom of Mn^{2+} deficiency is leaf interveinal chlorosis.

Sodium

Sodium (Na) is required by plants with C_4 and CAM photosynthesis (Chapter 10) for the regeneration of PEP, the molecule to which CO_2 is added to form the four-carbon compound oxaloacetate (OAA) that transports CO_2 between cells or cellular compartments (Taiz and Zeiger 2006). Some plants are natrophilic (Na-loving) and thrive only if sufficient Na is available. In these plants, which often grow in saline soils, Na^+ can function along with K^+ as an osmoticum in the vacuole to draw water into cells to support growth (Chapter 8).

Zinc

Zinc (Zn) is a component of numerous enzymes, notably those involved in the production of RNA from DNA (transcription). A Zn deficiency interferes with plant growth, and symptoms can vary from lack of stem elongation in dicots to interveinal chlorosis in monocots.

Copper

Copper (Cu), which can exist as Cu^+ or Cu^{2+}, is a component of enzymes involved in oxidation–reduction reactions, particularly in electron transport. An early symptom of Cu deficiency is production of dark green leaves with rolled margins.

Nickel

A deficiency of nickel (Ni) is rarely seen in nature. However, when a Ni deficiency was induced in plants involved in symbiotic nitrogen fixation, it resulted in the buildup of the nitrogen compound urea, causing necrosis (burn) of leaf tips.

Molybdenum

Molybdenum (Mo) is required for nitrogen metabolism, as a component of the enzyme that catalyzes the first step in the utilization of NO_3^-, and as a component of the enzyme complex that fixes atmospheric nitrogen (N_2) by reducing it to ammonia (NH_3). Deficiencies of Mo occur on some acidic soils and result in interveinal chlorosis of older leaves.

Fertilizer formulation

Commercial fertilizers often contain a balance of the three mineral macronutrients that are most often deficient in soils or media used for crop production: nitrogen, phosphorus, and potassium (N, P, and K). The percentage by dry weight of N in the fertilizer is the first of the three numbers describing the nutrient content on the label (the **N-P-K ratio**). The second and third numbers represent P and K, respectively, but their proportions are reported in the form of P_2O_5 and K_2O, regardless of the actual mineral form used in the fertilizer (Figure 7.12). To calculate the amount of these

Jack's Professional
20-20-20 General Purpose
Water Soluble Fertilizer
(For Continuous Liquid Feed Programs)

GUARANTEED ANALYSIS   Product # 77010

Total nitrogen (N) 20%
    3.83% ammoniacal nitrogen
    6.07% nitrate nitrogen
    10.10% urea nitrogen
Available phosphate (P_2O_5) 20%
Soluble potash (K_2O) 20%
Magnesium (Mg) Total 0.05%
    0.05% water soluble magnesium (Mg)
Boron (B) 0.0068%
Copper (Cu) 0.0036%
    0.0036 % chelated copper
Iron (Fe) 0.0500%
    0.0500% chelated iron
Manganese (Mn) 0.0250%
    0.0250% chelated manganese
Molybdenum (Mo) 0.0009%
Zinc (Zn) 0.0025%
    0.0025% chelated zinc

Derived from: urea, ammonium phosphate, potassium nitrate, magnesium sulfate, boric acid, iron EDTA, manganese EDTA, zinc EDTA, copper EDTA, ammonium molybdate

Potential Acidity: 555 lb. Calcium carbonate equivalent per ton

Manufactured by: J.R. Peters, Inc., 6656 Grant Way, Allentown, PA 18106
    Toll Free: 1-866-522-5752

Figure 7.12 Analysis label from Peters 20-20-20 fertilizer. Note that percentages of the macronutrients N, P, and K are given as percentages of N, P_2O_5, and K_2O. (Courtesy of R. J. Peters, Inc.)

elements that are really included in the fertilizer, it is necessary to know how much P by weight is in P_2O_5, and how much K by weight is in K_2O.

The elemental content of a fertilizer can be calculated by comparing the mass of each element to that of the respective molecule. The atomic mass of P is 31 and the atomic mass of oxygen (O) is 16. A molecule of P_2O_5 has two atoms of P and five atoms of O. Therefore, it has a total atomic mass of $(2 \times 31 = 62) + (5 \times 16 = 80) = 142$. The percentage of P in each P_2O_5 molecule is the mass of P in the molecule divided by the mass of the whole molecule: 62/142 or 44%. The elemental content of P in a 20-20-20 fertilizer is 20×0.44, or 8.8%.

The atomic mass of potassium is 39, so one molecule of K_2O has a mass of $(2 \times 39) + 16$, or 94. The percentage of K in K_2O is 78/94 or 83%. The K content of a 20-20-20 fertilizer is 20×0.83, or 16.6%.

Grasses and foliage houseplants have a relatively high N requirement for vegetative growth and require a fertilizer such as 27-15-12 (which is actually 27% N, 6.6% P, and 9.96% K). Too much N applied to plants grown for their flowers or fruit will encourage continued vegetative growth and reduce flowering; a low-N fertilizer (such as 10-30-20) applied after vegetative maturity will promote the development of flowers and reproductive organs. Root crops also require relatively less N as their food storage organs develop, and a higher K fertilizer (such as 15-11-29) is recommended to promote development of roots and tubers.

Chapter 8
Plant–water relations

Water is fundamental to the lives of plants. The nutrients plants extract from the soil are taken up as solutes in soil water. While plants are not mobile, they use the flexible growth of their root systems to seek out nutrient-rich regions of the soil, and some plant roots grow deep into the soil to reach moisture in otherwise dry surroundings (Chapter 3). The need for water is also driven by photosynthesis, which requires the uptake of CO_2 from the air surrounding leaves. When stomata open to allow CO_2 to diffuse in, water vapor escapes. As we saw in Chapter 5, plants have leaves adapted to be less susceptible to water loss, and as we will see in Chapter 10, variations on photosynthesis have evolved that are more workable in hot, dry environments. In this chapter, the fundamental roles of water in plants are discussed.

Uptake of water

Water is the most common molecule in plants, and is an excellent solvent. A **solvent** is a liquid capable of dissolving or dispersing other substances. The substances that become dissolved in water in plants include mineral ions such as potassium (K^+), sugars such as glucose and fructose, and amino acids, the building blocks of proteins. The substances that become dissolved are called **solutes**.

The movement of molecules from a location where they are in a high concentration to an area where they are in a lower concentration is called **diffusion** or passive transport (Figure 8.1). Diffusion occurs because of the natural random movement of molecules. Perfume spilled in one corner of a room can quickly be detected throughout the entire room because of diffusion.

Osmosis is a special type of diffusion (Figure 8.2). It is the diffusion of water through a **differentially permeable membrane**, such as a plant cell membrane that allows water through but not solutes. Cells must take up water for growth to occur, and herbaceous (nonwoody) plants must maintain

Diffusion

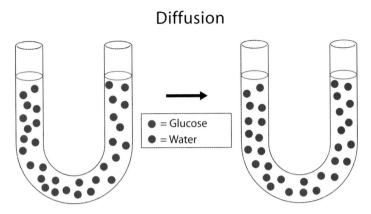

Figure 8.1 Diffusion is the redistribution of a solute such as glucose within a solvent such as water, from a region of high concentration to regions of lower concentration. In this case, a high concentration of glucose molecules (red) is redistributed by diffusion to create a lower concentration among the solvent water molecules (blue).

water pressure in their cells just to retain their shape. As we will see, they do this using osmosis. In osmosis, water moves from an area where water is in a high concentration (a dilute solution with few solutes and lots of water) to an area where water is in a low concentration (a concentrated solution higher in solutes and therefore lower in water).

Water and gases such as oxygen (O_2) and carbon dioxide (CO_2) can pass through plant cell lipid membranes easily, but solutes such as potassium and sucrose cannot move into cells through the differentially permeable plant cell membrane. Protein molecules, or in some cases protein complexes

Osmosis

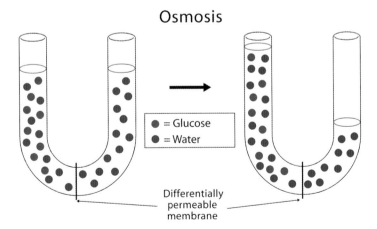

Figure 8.2 Osmosis drives the uptake of water into cells, creating turgor. The plasma membrane is a differentially permeable membrane that allows water movement but controls the movement of solutes such as potassium; therefore, a cell can regulate water uptake by regulating the uptake of a solute. If a cell has a higher concentration of a solute inside than outside (as the glucose molecules concentrated on the left side of the vessel), water is less concentrated inside the cell and will move across the plasma membrane to make its own concentration the same inside as outside the cell.

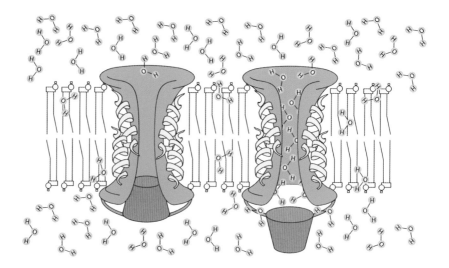

Figure 8.3 Aquaporins are channels in the plasma membranes of cells that allow rapid movement of water molecules. Aquaporins can close (left) or open (right) to regulate water movement.

associated with the plasma membrane, act as gateways to regulate the uptake of small molecules that cannot diffuse through the lipid bilayer. There are also channels in the plasma membrane specifically for the rapid movement of water called "**aquaporins**" (Figure 8.3).

Plant cells control water uptake via osmosis by accumulating molecules such as potassium in cells against a concentration gradient, by expending energy. This is called **active transport**, and the energy usually comes from the molecule ATP that is formed in cells by respiration (Figure 8.4).

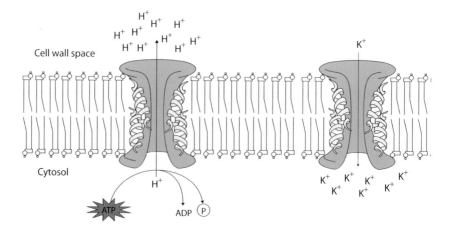

Figure 8.4 Nutrients and other solutes often have dedicated channels in the plasma membrane for inward and outward movement, and their uptake is powered by creation of a charge imbalance. A charge deficit is created using ATP to pump H^+ out of the cytosol after which K^+ ions flow in to replace H^+. Anions such as NO_3^- can flow into cells accompanied by returning H^+. Using the energy of ATP to promote movement into cells is termed "active uptake."

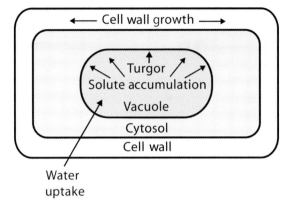

Figure 8.5 Solute accumulation in the vacuole is followed by the uptake of water. The resulting turgor pressure acts to expand the vacuole and cell wall. This would cause the cell wall to thin, but new cell wall structural compounds are added to the wall, resulting in growth of the cell wall and a reduction in turgor pressure. This cycle of cell wall deposition and water uptake is repeated continuously during cell expansion.

When a solute molecule like K^+ is accumulated to a higher concentration inside a cell than outside, water will diffuse by osmosis across the lipid bilayer or via aquaporins in the cell membrane in an attempt to equalize the concentration of water inside and outside the cell. Because plant cells are enclosed in a rigid cell wall, the volume of water that can be taken up is limited. Therefore, the **turgor pressure** generated by water in the vacuole of a mature cell causes the cytoplasm and plant cell membrane to be pressed against the cell wall. The presence of the rigid cell wall allows the cell to be **turgid** or to have a positive turgor pressure that offsets part of the negative solute potential drawing water in and allows cells to have a lower solute potential than the water in the apoplast (outside the cell). It is normal for a plant cell to be turgid or under pressure; when turgor pressure is reduced, the result is wilting. When grass leaves are subjected to drought, their large bulliform cells (Chapter 5) lose turgor, causing the cells to collapse and the leaf to roll.

In growing cells, cell walls are able to expand through the deposition of new cell wall material, and the active uptake of solutes followed by the uptake of water and the resulting turgor pressure causes enlargement of the cell wall (Figure 8.5). Growth is thought to occur by alternate cycles of cell wall synthesis followed by accumulation of solutes and therefore of water, which causes the cell to expand, or grow.

Movement of water in plants

Water potential is the force that causes water to flow from locations where it is relatively more pure or where it is under pressure to locations where water is under less pressure or where it has a higher solute content. The potential of pure water in an open beaker has been set at zero and adding solutes to water lowers water potential into negative numbers. The lower or more negative

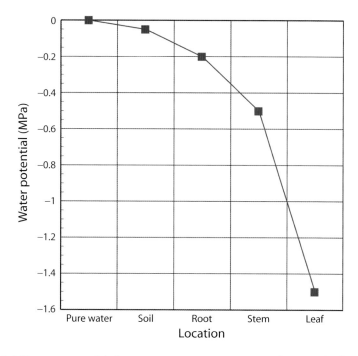

Figure 8.6 The water potential of pure water at room temperature and pressure has been set at zero. The addition of a solute reduces water potential, so water potential in soils and plants is always negative. Water will move along a gradient from a higher to a lower water potential, which can be envisioned as water flowing down a water potential gradient.

the measured water potential of a plant organ, such as a leaf or root, the more readily it will take up water. Just as water flows downhill, it flows down a water potential gradient (Figure 8.6). Water potential can be reported as a negative number in various units of pressure, such as atmospheres, bars, and inches of mercury. The unit **megapascal** (MPa) is preferred.

The water potential within a plant is always less than zero, because the liquid water in the vacuole and the cytosol contains solutes such as nutrient ions and sugars. The water potential in roots is either equal to or more negative than that of the soil solution surrounding the roots, or water would flow from the plant into the soil. Flow from roots into soil does occur in arid soils when very deep-rooted plants move water from deep, moist soil layers through dry soil into the shoot for photosynthesis. At night, when stomata close, the entire root and shoot equilibrate with the moist, deep soil and water can flow from the upper root system into dry, surface soil. This process is termed "hydraulic lift" (Figure 8.7).

A soil at "field capacity"—containing as much water as it can hold against gravity (Chapter 7)—has a water potential of about −0.03 MPa. The water potential of a well-watered soil surrounding the roots of a plant has a small but negative water potential, such as −0.05 MPa (Figure 8.8). These negative water potentials are generated by minerals dissolved in the soil solution that lower the water potential of the soil solution compared with pure water. The

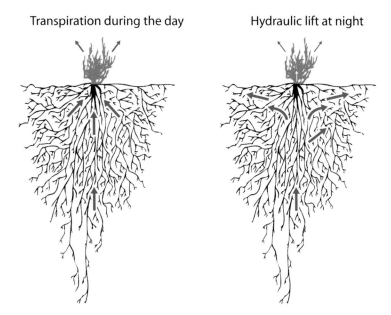

Figure 8.7 Deep-rooted plants can reach moist soil and support photosynthesis and the accompanying water loss by transpiration during the day. At night, stomata close but water uptake continues to rebalance root and shoot water potentials. Water in root systems of deep-rooted plant species can be lost to the dry surface soil, supplying water to neighboring plant species. (Figure from Caldwell 1988, used with permission.)

water potential of a saline soil solution can be significantly lower or more negative than −0.05.

In order to draw water out of the soil, the cells of the root must have a lower or more negative water potential than the soil. So the water potential of the root may be −0.2 MPa if the water potential of a well-watered soil is −0.05 MPa; this is low enough that water will flow from the soil into the root.

Living cells of the root outside the stele—in the epidermis and the cortex—are responsible for accumulating ions and thus water from the soil solution. Just as the endodermis acts as a gateway for the uptake of mineral nutrients, it also prevents ions from moving from the stele back into the cortex. Therefore, nutrient ions that have moved from the endodermis into the xylem parenchyma, the living cells surrounding the nonliving xylem elements and tracheids, are transported from the symplast into the apoplast of xylem parenchyma and into water-conducting xylem elements by ion-specific passive and active transport. The living xylem parenchyma cells inside the stele generate the osmotic potential needed to draw water into the xylem from the cortex of the root.

Transpiration

Osmosis is diffusion of liquid water into plant cells. The *loss* of water from leaves also occurs by diffusion, but in this case, by the diffusion of water vapor from the air spaces inside the leaf to the air that surrounds the leaf, a

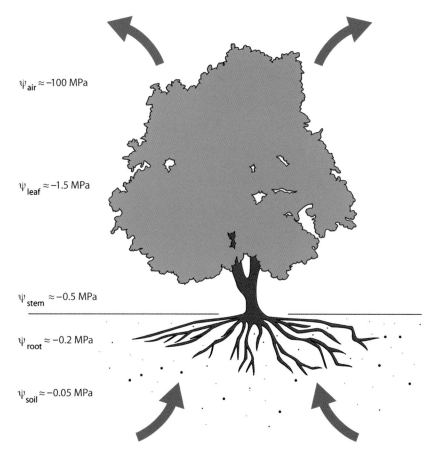

Figure 8.8 Water is transpired from the saturated inner surfaces of leaves to dry surrounding air when stomata open for photosynthesis; it is replaced by water absorbed from the soil. Uptake of water from the soil is driven by a water potential gradient: root water potential must be lower than soil water potential to attract water, and progressively lower water potentials exist within the plant from root to leaves.

process called **transpiration**. Inside the air spaces of the leaf, the air is saturated, or holding all the water vapor it can at a given temperature. Outside the leaf, the air is much drier than inside unless the relative humidity outside is also saturated because it is raining. Therefore, water vapor diffuses from inside the leaf, where it is in high concentration, to outside the leaf, where water vapor is in a lower concentration.

In leaves, the water potential in the air spaces inside the leaf where water vapor is concentrated is high relative to the surrounding air, which is much drier because the molecules of water are more dispersed in the outside air. Therefore, the air surrounding a leaf has a lower, or more negative, water potential than the leaf. Water vapor will move rapidly from the leaf into the surrounding air when stomata are open.

The water potential of the air surrounding a leaf will be a function of the water content of the air and the temperature (warm air can hold more water than cold air), but will probably be a large negative number such as −100 MPa. If the relative humidity is very high, the water potential of the air may be reduced to 1/10th this number, or −10 MPa, still much lower than the water potential inside the leaf.

As leaves lose water to the air, water in the leaves is replaced by water from the stem, which will have a water potential between that of the roots and the leaves, in the range of −0.5 MPa (Figure 8.8).

The water potential of the leaf is not only a function of the humidity of the air in the air spaces between the cells, but also of the cells of the leaf that are using solute concentrations and osmosis to remain turgid, and so is usually on the order of −1.5 MPa, still much less negative than the outside air.

Fresh water—a finite resource

Even though about 70% of earth's surface is ocean, the fresh water required for our own consumption or for plant growth is a finite resource. Fresh (not saline) water, including the water in ice caps, glaciers, groundwater, and soil moisture, represents only 2.5% of the water on earth (Pimentel et al. 1997). About a quarter of this fresh water is available for our use as groundwater or in lakes and streams.

Pimentel et al. (1997) distinguish between water use, where at least some of the water can be recovered and reused, and water consumption, where water is lost back to the atmosphere and can be returned only through precipitation. Humans consume less than 2 liters of water per day, but personal water use in the United States is about 400 liters per day. Total per capita use of fresh water in the United States is about 5,100 liters per day when food and industrial production are factored in.

Consumption of fresh water by plants for transpiration, which is required for photosynthesis, ranges from 500 to 2,000 liters (or kg) of water per kg of yield. A gallon of water weighs a little more than 8 pounds, so a pound of crop yield requires between 60 and 240 gallons of water to produce. A crop of maize on productive soil can consist of as many as 75,000 plants per hectare (30,000 plants per acre), and over a 4-month growing season the plants on each hectare of soil will transpire 4 million liters, which is about 50 liters (13 gallons) of water per plant. It is estimated that 87% of fresh water is used for agricultural (crop and livestock) production in the United States (Pimentel et al. 1997).

When the water for plant growth comes from rainfall, the crop only benefits from the water plants can remove from the root zone. The rate of infiltration of water into the soil depends on the type of soil and whether there is a groundcover or bare soil. Water that exceeds the infiltration rate (e.g., a heavy rain) will run off, carrying soil and nutrients. As discussed in Chapter 7, water in a saturated soil can leach through the root zone, carrying nutrients such as nitrate (NO_3^-) into the groundwater.

When precipitation is insufficient or out of sync with crop production, irrigation can be used to supply water for crop growth. The costs associated with irrigation include irrigation equipment and maintenance of reservoirs and canals as well as the cost of the energy required to pump or deliver water. Therefore, irrigated land is generally used for higher value crops. The water used for irrigation contains varying concentrations of dissolved salts which become concentrated by transpiration of pure water from the plant. Fertilizer salts used for crop growth may be added, so irrigated soils must be managed carefully to prevent salinization.

The Ogallala aquifer (Figure 8.9), which underlies much of the High Plains, consists of sand, gravel, and porous rock saturated with water (Nativ and Smith 1987). It

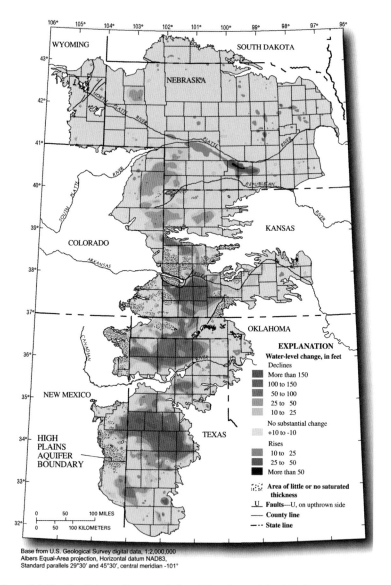

Figure 8.9 The Ogallala aquifer extends from Wyoming and South Dakota to New Mexico and Texas. It has been used for agricultural irrigation since the 1940s. This figure shows the change in water level in the Ogallala aquifer during the period from 1950 through 2003. Yellows and reds indicate decrease in water levels; greens and blues indicate a rise. (Figure from USGS; McGuire 2004.)

supplies about 30% of all the groundwater used for irrigation in the United States (Rosenberg et al. 1999). Use of this aquifer for irrigation has increased since the 1940s; annual withdrawal of groundwater was estimated at 4.9 billion cubic meters of water in 1949 and increased to a high of 23.4 billion cubic meters by 1974. However, recharge of this aquifer is exceedingly slow, and water levels in the aquifer have declined an average of 3.84 m (12.6 ft) from 1950 through 2003 (McGuire 2004). However, in parts of Texas, Oklahoma, and southwestern Kansas, the water level has declined more than 30 m (100 ft). In Figure 8.9, reds and yellows represent decline in water levels, while greens and blues represent locations where water levels in the aquifer have risen. The Ogallala aquifer is quite a striking example of the finite nature of the water that sustains us and has become a symbol of the need for the use of highly efficient (but also very expensive) irrigation technologies.

A "**crop coefficient**" is a percentage used to describe the amount of soil water that will be transpired by a particular crop compared to a reference crop, and this coefficient will vary depending on the crop's stage of development as well as the potential evapotranspiration (ET), which is the evaporation from the soil surface and transpiration from plant leaves. The crop coefficient of a perennial crop such as alfalfa that is harvested several times each growing season decreases sharply after harvest, when all the shoots are removed (Figure 8.10), and then rises as the crop regrows, becoming constant after the canopy is fully developed and transpiration is maximal. Change in the crop coefficient is directly related to increase or decrease in leaf area, because transpiration from leaf surfaces will occur when photosynthesis occurs.

The opening of stomata and transpiration occur because the plant requires CO_2 from the surrounding air for photosynthesis. For most plants, stomata open in the morning after the plant has been exposed to the sun long enough to use up most of the CO_2 inside the leaf, and stomata close at night when there is no longer light for photosynthesis and therefore no longer a need for CO_2 influx (Figure 8.11a). The loss of water is incidental to the need for uptake of CO_2. However, in some xerophytic plants adapted to carry out photosynthesis in very dry climates (Figure 8.11b), stomata open for CO_2

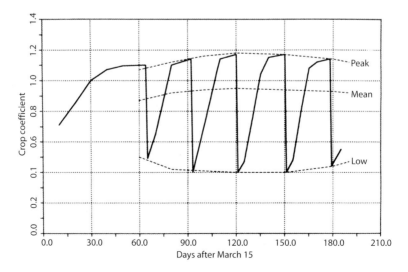

Figure 8.10 A crop coefficient describes a crop's transpiration and evaporation from a given land area. Transpiration is a function of leaf surface area that changes with stage of growth. Initial spring growth and regrowth cycles following each harvest of an alfalfa hay crop can be seen in its pattern of water use. (Figure from James et al. 1982, used with permission.)

uptake at night when transpiration is greatly reduced. The CO_2 is stored as malic acid in the vacuole, then released and used to carry out photosynthesis during the day (CAM photosynthesis; Chapter 10).

At night, stomata close because there is no need for CO_2 for photosynthesis, so transpiration will stop and the water potential of leaves will become less negative as water continues to flow from the roots to the leaves. Leaf water potential becomes equal to that of the soil, and in fact the water potential of the soil is often measured by taking a "predawn" (before photosynthesis begins) water potential of leaves.

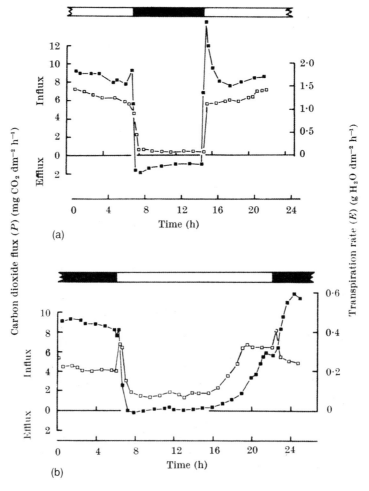

Figure 8.11 Transpiration occurs when stomata open for the uptake of CO_2 for photosynthesis. In both figures, the dark sections of the bar indicate the dark period. CO_2 influx and transpiration occur during the light period in the C_3 species tobacco (*Nicotiana glutinosa* (a)) and during the dark period for the CAM photosynthesis species century plant (*Agave americana* (b)). CAM plants are succulents that open their stomata for CO_2 uptake at night, storing CO_2 as malic acid for use during the subsequent light period (Chapter 10). (Figures from Neales *et al.* 1968, used with permission.)

Drought

When plant are subjected to drought, the soil dries and the water potential of the soil decreases, so root water potential must also decrease if the plant is to continue to remove water from the soil to meet transpiration demands. At night, leaf water potential can only rise to the water potential of the soil. If drought continues, a soil water potential is reached from which plants cannot take up water, and this is termed the "**permanent wilting point.**" This occurs at a soil water potential of about −15 bars or −1.5 MPa (Figure 8.12).

When a plant is drought stressed, the control of stomatal opening by the need for CO_2 is overridden by the need to preserve a favorable water status for the plant, and this response to drought involves the plant growth regulator abscisic acid (ABA; Chapter 13). For stomata to open, guard cells take up potassium and other charged molecules that drive water uptake, and as the guard cells become turgid the stoma opens. ABA causes guard cells to close in daylight by causing them to secrete the ions that provide them with turgidity, which causes the guard cells to become flaccid, and the stoma closes (Figure 8.13).

Because growth depends directly on water uptake, when a plant experiences drought, growth is quickly inhibited. When stomata close during the day in response to drought, photosynthesis continues for a while but

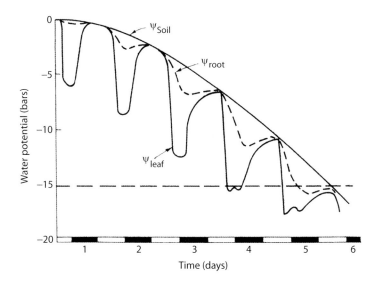

Figure 8.12 The water potentials (Ψ) of the soil (upper line), the root (dashed line), and the leaf (lower line) of a plant in a progressively drier soil. Dark sections of the bar indicate night periods. For water uptake to continue, the decrease in Ψ_{root} must be more than that of Ψ_{soil}. Ψ_{leaf} decreases during the light period as stomata open for photosynthesis and transpiration occurs. It stops decreasing when absorption of water from the soil equals transpiration, and then increases to the level of the Ψ_{soil} when stomata are closed during dark periods. When the soil reaches the permanent wilting point indicated by the horizontal dashed line (−15 bars or −1.5 MPa), the plant is unable to reduce its own water potential further and can no longer take up water. (Figure 9.1, p. 276 in Slatyer 1967, used with permission.)

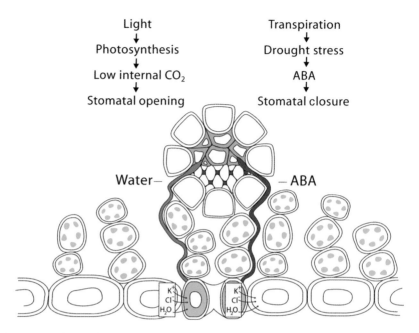

Figure 8.13 Photosynthesis can occur using internal CO_2 when a leaf is exposed to light; stomata open (guard cells become turgid) in response to a low internal concentration of CO_2 by taking up solutes such as potassium (K^+) and chloride (Cl^-) followed by water uptake. When a plant experiences drought, stomata are regulated by ABA and other factors to close guard cells by secreting K^+ and Cl^- to reduce turgor, even though internal CO_2 concentrations continue to decrease.

eventually is inhibited by a buildup of the by-products of the light (photochemical) reactions of photosynthesis that cannot be used because there is no CO_2 uptake to supply the dark (biochemical) reactions of photosynthesis (Chapter 10). These compounds are highly reactive and can damage the membranes that are used in photosynthesis, and therefore photosynthesis may not recover as quickly as growth after drought stress is relieved. Growth can often recover quickly from a short drought because cell wall synthesis continues for some time in the absence of the water needed to fill expanding cells. When water uptake resumes, cells may only need to take up water to expand to full size. This spurt of growth in plants following a mild drought is called "compensatory growth" (Figure 8.14).

Water movement in stems

The movement of water upward through a plant is possible because of the differences in water potential between the leaves at the top of the plant and the roots in the soil. Considering the force of gravity on the trunks of tall trees, the upward movement of water is also dependent on the nature of the water molecule itself.

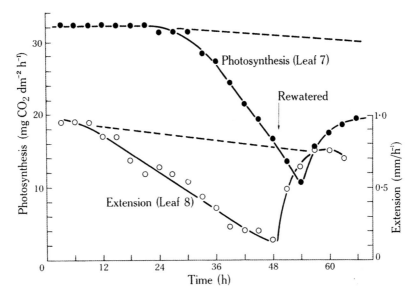

Figure 8.14 Recovery of growth following a total of 5 days of drought is more complete than recovery of photosynthesis. Dashed lines represent leaves of control plants of darnel ryegrass (*Lolium temulentum*). Extension of Leaf 8 began to be affected 3 days after imposition of drought but recovered to its original growth rate within 12 hours of rewatering. Photosynthesis rate of the older Leaf 7 was not affected for the first 4 days of drought (at 24 hours on this graph), but only returned to about half its original rate a day after rewatering. The products of the light reactions of photosynthesis (ATP and NADPH) accumulate when internal CO_2 is low because stomata are closed, and can damage the photosynthetic apparatus. (Figure from Wardlaw 1969, used with permission of CSIRO Publishing, www.publish.csiro.au/journals/.)

A water molecule is made up of one atom of oxygen and two atoms of hydrogen (H_2O). Water molecules are not charged, but the oxygen atom has a stronger attraction for the electrons of the molecule than the two hydrogen atoms, so the oxygen atom carries a partial negative charge and the hydrogen atoms carry a partial positive charge within each water molecule (Figure 8.15). The unequal charge distribution on a water molecule makes it **polar** (carrying equal and opposite charges separated by a small distance). This polarity causes the partially positive regions of water molecules to be attracted to the partially negative regions of other water molecules, giving

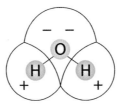

Figure 8.15 The oxygen in a water molecule has a stronger attraction for electrons than the hydrogen atoms, so oxygen carries a partial negative charge and the hydrogens carry a partial positive charge.

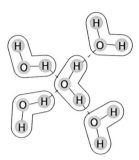

Figure 8.16 Hydrogen bonds form between the oxygen atom of one molecule of water and the hydrogen atoms of the water molecules around it due to the polarity of water molecules (Figure 8.15). Each hydrogen bond is relatively weak but many hydrogen bonds provide considerable structure to liquid water.

water an unusually ordered structure for a liquid and a unique set of chemical characteristics. The strong attraction of water molecules for each other is termed **cohesion** (Figure 8.16).

The surface tension of water is caused by water molecules at the surface of a droplet being pulled in toward the rest of the molecules in the water droplet, and is the reason water forms spherical droplets (Figure 8.17). Surface tension even in flowing water can support the weight of small insects. When a surfactant or spreader is added to an aqueous solution, it interferes with cohesion and causes water and the solutes it contains to spread more evenly over the surface of plants. Spreaders are used with the application of some compounds to leaves to aid in absorption.

A large amount of heat is required to separate water molecules and change liquid water into water vapor; this is why evaporation (and transpiration) provides significant cooling (Figure 8.18). It is also why the temperature of water is very stable and can absorb a great deal of heat before its temperature rises. This helps to provide thermal stability for plants, which are typically 75–80% water.

Cohesion also gives water **tensile strength**, which is the tension (pull) a continuous column of a material can withstand before it breaks. Water columns

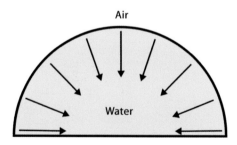

Figure 8.17 Surface tension results from the inward pull of the hydrogen bonds between water molecules at the surface and those in the rest of the droplet.

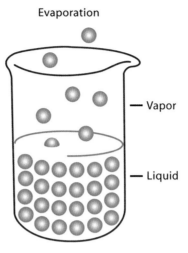

Figure 8.18 When a water molecule evaporates, the hydrogen bonds between that molecule and the surrounding water molecules must be broken. Evaporative cooling results from the absorption of heat (energy) required for this process.

Figure 8.19 The cohesion that results from hydrogen bonding among water molecules provides remarkable tensile strength that allows a column of water even in a tall tree to remain unbroken under tension caused by transpiration.

in stems are being pulled up by transpiration losses from leaves and are therefore under considerable tension (Figure 8.19). Their length is the distance from the leaf to the root tip, because xylem water is continuous within plants. It has been determined that a thin water column in a xylary element can be pulled to the top of a tree by a tension of -2 MPa, but could withstand a tension of -30 MPa without breaking.

Chapter 9

Macromolecules and enzyme activity

Organic chemistry is the study of the carbon-based molecules that are comprised in living organisms, and biochemistry is the study of the interaction of these **organic molecules** in enzyme reactions and biochemical pathways that allow living organisms to function. In this chapter, protein synthesis and enzyme activity are discussed along with the structures of carbohydrates, lipids, nucleic acids, and proteins, which are the organic molecules synthesized in plants. In the following chapters, we examine photosynthesis and respiration, the critical biochemical pathways that supply raw materials and energy for the synthesis of organic molecules on which all living creatures depend.

Chemical bonds

Hydrogen bonds (Figure 9.1) occur when the orientation of molecules is strongly influenced by the polarity of atoms associated with hydrogen; often these are oxygen-containing molecules. The orderliness of water is due to the attraction of the oxygen atom of one water molecule for the hydrogen atoms of other water molecules. Hydrogen bonds are important in the self-assembly of cellulose molecules into microfibrils and hold the strands of nucleic acids together into the double helix of DNA.

Ionic bonds (Figure 9.2) form when one atom, such as sodium, gives up an electron to another atom, such as chloride. In a solid like table salt (NaCl), the sodium and chloride remain physically close to each other because they need their original number of electrons for balance. In solution, as in irrigation water or the soil solution, we find the separated ions (Na^+ and Cl^-).

Covalent bonds (Figure 9.3) occur when two atoms share a pair of electrons. Most atoms are less stable on their own than when combined with atoms that have a complementary number of electrons. For example, oxygen needs two more electrons to have a completely stable number (it has eight

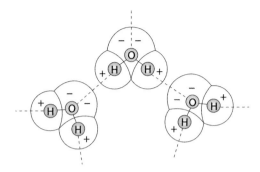

Figure 9.1 Hydrogen bonds: atoms such as oxygen or nitrogen have a stronger attraction for electrons than hydrogen. Therefore, hydrogen atoms carry a partial positive charge when they are associated with these atoms, and this can lead to attractions between the partial positive charge on hydrogen in one molecule and the partial negative charge on oxygen or nitrogen in another molecule.

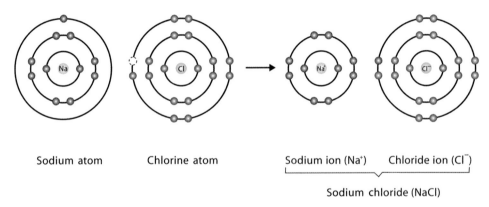

Figure 9.2 Ionic bonds form when an electron is donated by one atom to another. Since electrons carry a negative charge, the atom losing the electron becomes positively charged (a cation) and the atom gaining the electron becomes negatively charged (an anion). These opposite charges cause the two atoms to remain attracted to each other.

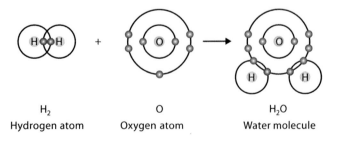

Figure 9.3 In covalent bonds, two atoms share one or more electron pairs. Molecules are composed of atoms joined by covalent bonds.

electrons and needs ten), and hydrogen atoms need two electrons to be stable but only have one. When one oxygen atom bonds with two hydrogen atoms, the oxygen can add the two electrons from the hydrogens to its outermost shell and share two of its original electrons with the two hydrogen atoms, so all three atoms are stable when bonded together to form water (H_2O).

Macromolecules

Dehydration reactions

Many organic molecules are made up of subunits or smaller monomers, which may be identical to each other, or fall within a class of similar molecules. A common reaction in the formation of such organic polymers is the **dehydration reaction** (Figure 9.4), which results in the loss of the components of water from two monomers as they become joined. The opposite reaction, the addition of water to break a polymer into its component monomers, is termed "hydrolysis."

Carbohydrates—structure and energy storage

Carbohydrates are sugars and polymers of sugars containing carbon, hydrogen, and oxygen, usually in the proportions CH_2O. The names of many

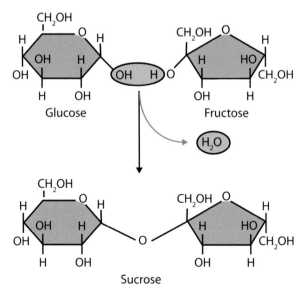

Figure 9.4 Many organic polymers are formed by dehydration reactions in which hydrogen (H^+) is removed from one monomer and a hydroxyl (OH^-) is removed from another monomer, producing a molecule of water and causing the two monomers to join each other.

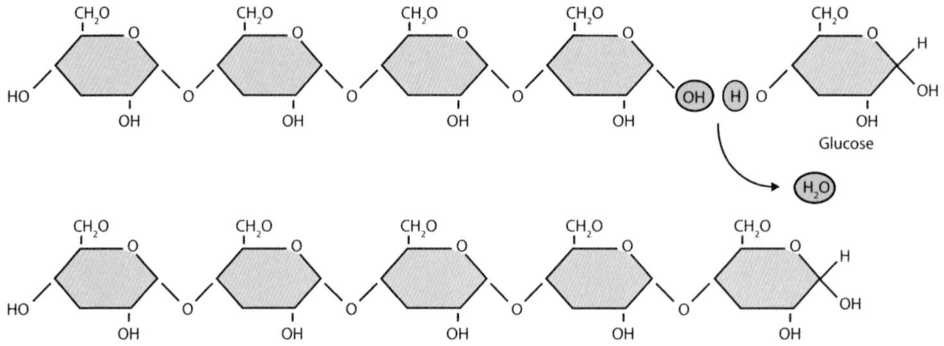

Figure 9.5 Starch is a polymer of glucose molecules which is formed by successive dehydration reactions that add glucose monomers to the growing starch chain.

carbohydrates end in "-ose," like glucose, sucrose, and cellulose. Carbohydrate polymers are synthesized by dehydration reactions (Figure 9.5). Some carbohydrates function in energy storage and others are used as structural molecules, as in cell walls. Cellulose and starch are both linear molecules of glucose, but while the linkages between glucose molecules in starch can be broken by mammalian digestive enzymes and therefore used as food, cellulose can only be digested by ruminants (and termites) using enzymes produced by microbes that inhabit their guts (Figure 1.3).

Lipids—membrane components and energy storage

Lipids are the group of organic compounds that include fats and membrane lipids such as phospholipids found in the plasma membrane. Fats are built by the attachment of three fatty acids to a three-carbon glycerol molecule (triglycerides) (Figure 9.6). Phospholipids are similar to triglycerides, except that one of the three fatty acids attached to the glycerol molecule is replaced by a phosphate group, the hydrophilic "head" of the phospholipids.

Fatty acids are long chains of carbon and hydrogen atoms. In any organic molecule, carbon atoms can form four bonds with other atoms. Each internal carbon in fatty acids uses two of these sites to bond with the carbons on either side. If a fatty acid is "saturated," then the other two sites on all but the end carbon atoms are occupied by hydrogen atoms (Figure 9.7). Saturated fats usually come from animals and are solid at room temperature. Examples are butter and lard. When a fatty acid is "unsaturated," it means that one or more neighboring carbons are double bonded to each other and can link to only one hydrogen atom each. Unsaturated fats, usually from plants, are liquid at room temperature. It is not clear why, but unsaturated fats, and especially polyunsaturated fats such as olive oil, with many double bonds between carbon atoms, do not cause deposits of fat in blood vessels, and are therefore better for human health.

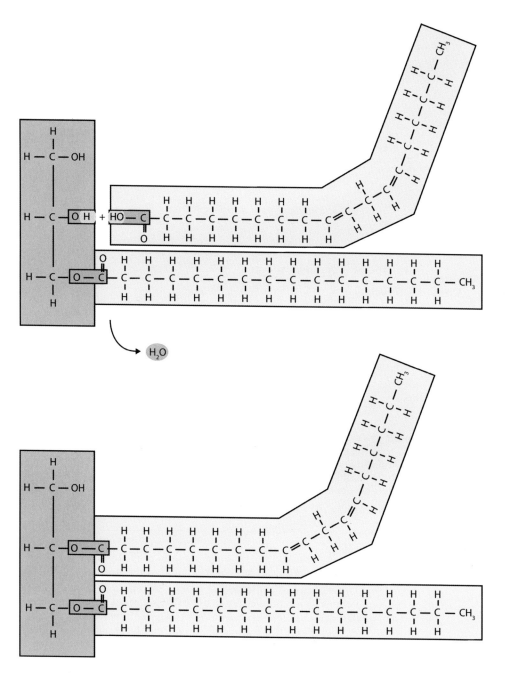

Figure 9.6 Fats are formed by dehydration reactions that add fatty acids to a three-carbon glycerol molecule.

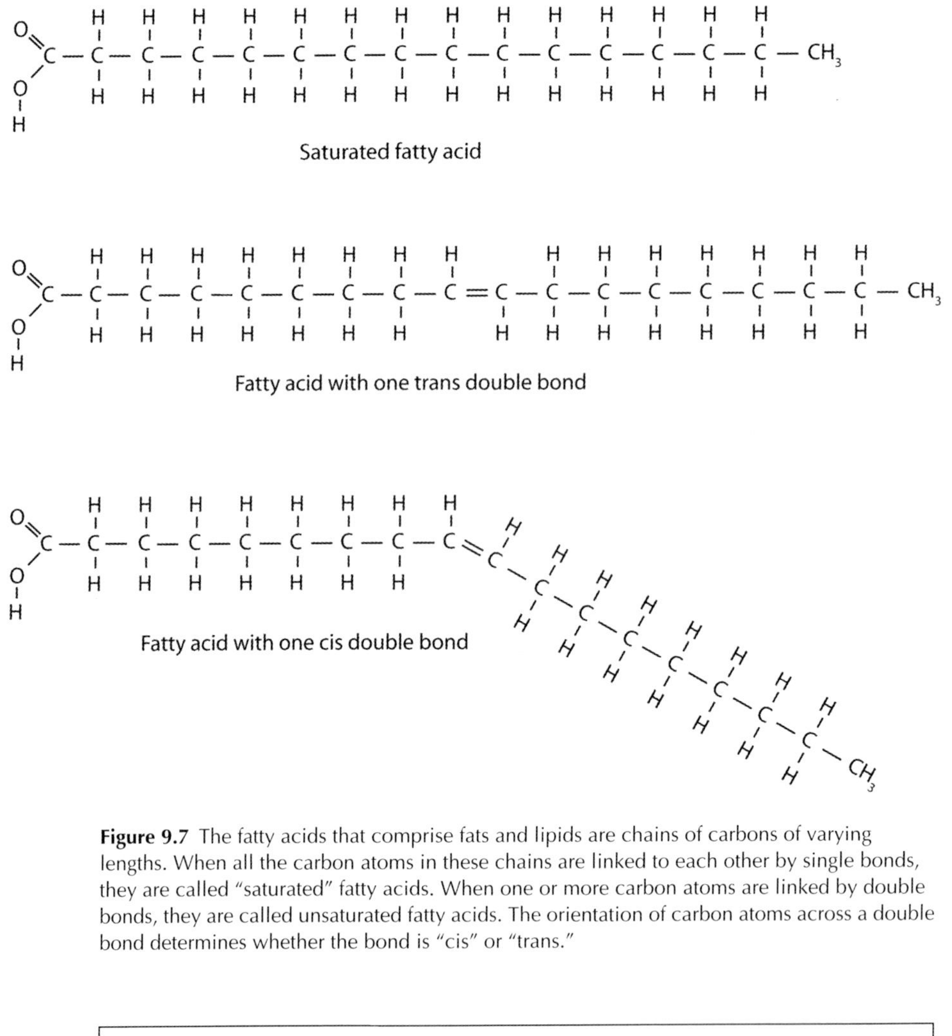

Figure 9.7 The fatty acids that comprise fats and lipids are chains of carbons of varying lengths. When all the carbon atoms in these chains are linked to each other by single bonds, they are called "saturated" fatty acids. When one or more carbon atoms are linked by double bonds, they are called unsaturated fatty acids. The orientation of carbon atoms across a double bond determines whether the bond is "cis" or "trans."

Trans fatty acids

Fatty acids are carbon chains with a carboxyl group (–COOH) at one end and a methyl group (–CH$_3$) at the other (Figure 9.7). Attachment of the fatty acid to glycerol occurs by a dehydration reaction in which the –OH of the carboxyl group is removed and the carboxyl oxygen becomes bonded to glycerol which contributes a H$^+$ to the dehydration reaction (Figure 9.6). Fatty acids are described by the number of carbon atoms they contain and the number of double bonds, if any. The fatty acids found in membrane lipids in plants include saturated fatty acids such as stearic acid (18:0), monounsaturated fatty acids such as oleic acid (18:1), and the polyunsaturated fatty acids such as linoleic acid (18:2) and α-linolenic acid (18:3).

The double bonds in fatty acids almost always have a cis conformation in which the carbon atoms are on the same side of the double bond (i.e., side-by-side), which creates a bend in the chain (Figure 9.7). The more cis bonds in a fatty acid, the more pronounced the curvature of the molecule. In contrast, in double bonds that have a trans conformation, the carbon atoms are on opposite sides of the double bond, and the molecule remains straight, like a saturated fatty acid. In nature, the milk and meat of ruminant livestock contain trace amounts of trans fatty acids, but trans fatty acids are not formed naturally in other foods.

In food industry, hydrogenation or reduction of fatty acids converts unsaturated fatty acids to saturated fatty acids by adding H atoms to existing double bonds. Saturated fatty acids increases the melting point of vegetable oils. Partial hydrogenation converts only some of the double bonds to single bonds, leaving one or more double bonds. The interaction of flour with the resulting semisolid fats in baking results in a more desirable texture than if oils were used, and partially hydrogenated fats resist oxidation and therefore have a longer shelf life than products made with oils.

A side effect of partial hydrogenation of vegetable fatty acids is that about two-thirds of the remaining cis double bonds are converted to trans double bonds. Trans fatty acids have become ubiquitous—they are found in pastries and used for frying in restaurants, and most notoriously were used in margarines that replaced butter because vegetable fats were considered more healthful. Trans fatty acids consumed in the diet increase the concentration of the undesirable LDL cholesterol and decrease the concentration of desirable HDL cholesterol. Trans fatty acids in the diet increase the risk of type II diabetes, and are also incorporated into cellular membranes in place of naturally occurring fatty acids, where their function is abnormal (Shaw 2004). Currently, the nutrition labels on packaged foods only require acknowledgment of trans fatty acids in foods if the content exceeds half a gram per serving, so a significant amount of these substances can be consumed unknowingly. Since relatively small amounts are not identified on food labels, the only way to determine that a product is free of trans fatty acids is to read the list of ingredients. If ingredients include "partially hydrogenated" vegetable oils, "shortening," or "margarine" then trans fatty acids are present.

Nucleic acids—storage and use of the genetic code

DNA and RNA are **nucleic acids**. These molecules are used to store the genetic code (as DNA in chromosomes) and to copy DNA (as messenger or mRNA) and synthesize enzyme proteins (as both ribosomal and transfer or tRNA). Chloroplasts and mitochondria also contain DNA, but is not segregated within a nuclear envelope. Nucleic acids are polymers of nucleotide subunits, and each **nucleotide** consists of a base, a sugar, and a phosphate group (Figure 9.8). Bases are heterocyclic rings (rings made of more than one element such as carbon and nitrogen) that vary from one nucleotide to another. In nucleic acid polymers, covalent bonds are formed between the phosphate of one nucleotide and the sugar of the next (Figure 9.9). In comparing DNA and RNA, the phosphates are the same, the sugars are very similar, and three of the four bases are the same. The sugar in RNA is ribose (RNA is *ribo*nucleic acid), and the sugar in DNA is deoxyribose (DNA is *deoxyribo*nucleic acid). The base is one of four in both DNA and RNA, and three of the four are the same, but thymine (T) in DNA is replaced by uracil (U) in RNA (Figure 9.8). As we will see (Chapter 11), ATP, used as an energy source, is very similar to the nucleotide adenine—it just has two additional

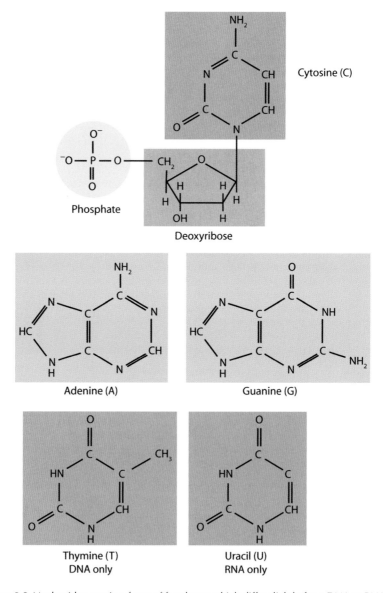

Figure 9.8 Nucleotides consist of one of four bases which differ slightly from DNA to RNA: a sugar—either ribose or deoxyribose (shown)—and a phosphate. Cytosine, adenine, and guanine are the bases found in both DNA and RNA; uracil is found in RNA, while thymine is found in DNA. Nucleotide monomers are the subunits of DNA and RNA.

phosphate groups. DNA consists of two chains of nucleotides that are linked by hydrogen bonds between pairs of bases and twisted into a double helix (Figure 9.9). One strand is copied; the other is complementary. The complementary strand runs in the opposite direction and contains a cytosine (C) opposite each guanine (G), a G opposite each C, a thymine (T) opposite each adenine (A), and an A opposite each T.

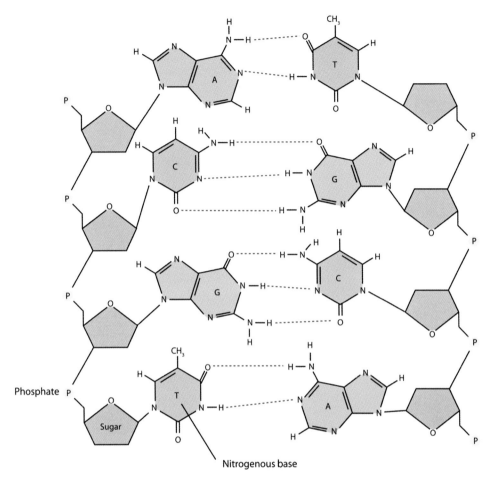

Figure 9.9 The double helix of DNA consists of two strands running in opposite directions held together by hydrogen bonds between cytosine and guanine or adenine and thymine. The backbone of each strand consists of the phosphate and deoxyribose groups of the nucleotides, with a base attached to each deoxyribose sugar.

Proteins—enzymes carry out the work of the cell

Plants function through the activity of thousands of enzymes in each cell, all of which are soluble proteins. **Proteins** are linear chains of amino acids. There are also nonsoluble or structural proteins that are found in cell walls along with structural carbohydrates. Proteins are abundant and make up 50% or more of the dry weight of living organisms. Seeds may contain storage proteins that act as a source of amino acids during germination.

Amino acids, the subunits of proteins, are composed of a central carbon (C) atom with bonds to a hydrogen atom (H), an amino group ($-NH_2$), a carboxyl group ($-COOH$), and a side chain with variable properties (Figure 9.10). Proteins are long chains of amino acids linked together with peptide

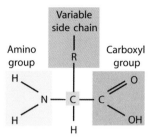

Figure 9.10 Amino acids consist of a central carbon atom to which are bonded a H, an amino (–NH$_2$) group, a carboxyl (–COOH) group, and a side chain (–R) that gives unique properties to each amino acid.

bonds, formed by a dehydration reaction when the carboxyl group of one amino acid is linked to the amino group of the next (Figure 9.11). There are 20 different amino acids used as the building blocks of proteins, but an enzyme may have 100 or 200 amino acids that are combined to give the enzyme unique properties, such as a hydrophobic region to orient the enzyme in the lipid bilayer of the plasma membrane (Figure 9.12). The sequence of amino acids making up the enzyme is determined by the gene (DNA sequence) that encodes for it. Of the 20 amino acids, those that the human body cannot make are called **essential amino acids** because they must come from the diet. Not all plants contain all essential amino acids, so it is important for humans who limit their diets to plants to consume products from a variety of plant species that together contain all the essential amino acids.

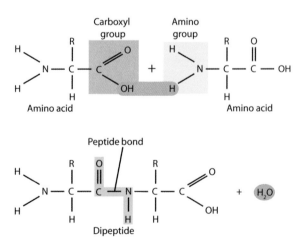

Figure 9.11 Proteins are polymers of amino acids joined by peptide bonds. Peptide bonds are formed by dehydration reactions that remove an OH from the carboxyl carbon and an H from the amino group, forming water. Proteins are linear chains in which the carboxyl groups of one amino acid is joined to the amino group of the next.

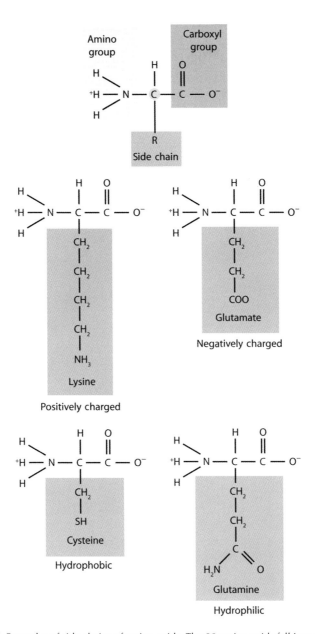

Figure 9.12 Examples of side chains of amino acids. The 20 amino acids fall into groups that are either positively or negatively charged, hydrophobic, or hydrophilic.

Protein synthesis

Three types of RNA are needed for protein synthesis:

1. Ribosomal RNA
2. Messenger RNA (mRNA)
3. Transfer RNA (tRNA)

All three are made in the nucleus, and exported to the cytosol, but only the mRNA is specific for the synthesis of a particular protein.

Ribosomes are composed of two subunits of ribosomal RNA that together comprise the physical location for protein synthesis. Some ribosomes are free in the cytosol, and others—the site of synthesis of proteins that will be excreted from the cell—are bound to the endoplasmic reticulum (ER).

mRNA is a copy of a gene, and dictates the order of amino acids that will comprise the protein. To synthesize mRNA, the DNA double helix is unwound, and a strand of RNA complementary to the DNA of the gene is created. This process is termed "transcription."

The mRNA is processed to delete unneeded sequences and to add a cap and a tail, and is then exported to the cytosol. It becomes sandwiched between the two subunits of ribosomal RNA, where the mRNA acts as a template for protein synthesis (Figure 9.13). Every set of three bases along the mRNA molecule specifies a single amino acid. tRNA molecules, also located in the cytosol, contain complementary combinations of three RNA bases. Furthermore, each of these tRNA molecules is specific for 1 of the 20 amino acids. The tRNA specified by the first three mRNA bases delivers the correct amino acid to the ribosome. The next tRNA that is specified by the next three-base mRNA sequence delivers its amino acid, which becomes attached to the first amino acid by a peptide bond, and the first tRNA is released. This process continues and the ribosome travels along the molecule of mRNA attaching amino acids to the elongating protein. This process is called "translation."

For proteins that will be exported, the ribosome is attached to the ER at a pore, and the elongating protein is fed into the lumen of the ER as it is synthesized. After processing in the lumen of the ER, such as the addition of a side chain (a sort of biological luggage tag), the protein is shuttled to the Golgi, and is eventually secreted (Figure 1.12).

The sequence of amino acids defines the **primary structure** of a protein (Figure 9.14a). The twisting and folding of the linear sequences of amino acids that results from interactions between hydrogen and oxygen atoms and hydrogen and nitrogen atoms results in the **secondary structure** of a protein (Figure 9.14b). The **tertiary structure** (Figure 9.14c) results from interactions among the "R" or unique side groups of the amino acids. This overall, three-dimensional structure is stabilized by disulfide bridges, a covalent bond between the sulfur atoms in two cysteine amino acids. An example of these disulfide bridges are the bonds in the keratin of hair that are first broken and then reformed when hair is permed to change its shape. When a protein consists of more than one chain of amino acids, the assembly of these subunits into a protein complex constitutes its **quaternary structure** (Figure 9.14d). Rubisco, the central enzyme of photosynthesis, consists of eight small and eight large subunits.

Enzymes

All enzymes are proteins, although all proteins are not enzymes. **Enzymes** are needed for the biosynthesis of all other organic molecules. A completed

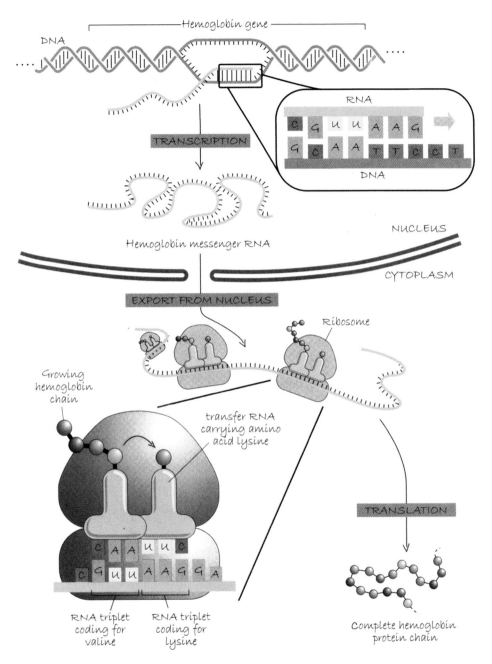

Figure 9.13 The synthesis of proteins occurs at ribosomes, either free in the cytosol or attached to the endoplasmic reticulum. A messenger RNA (mRNA), copy of the DNA of a gene, is formed in the nucleus (transcription). The mRNA is processed and exported from the nucleus and becomes attached to a ribosome. Every group of three RNA bases specify an amino acid, and transfer RNA (tRNA) molecules, complementary to the mRNA and also specific to a particular amino acid, deliver the correct amino acids to the ribosome. When a new amino acid is attached by a peptide bond to the elongating protein, the preceding tRNA is released and the ribosome progresses along the mRNA strand (translation). (Courtesy of Keith Roberts.)

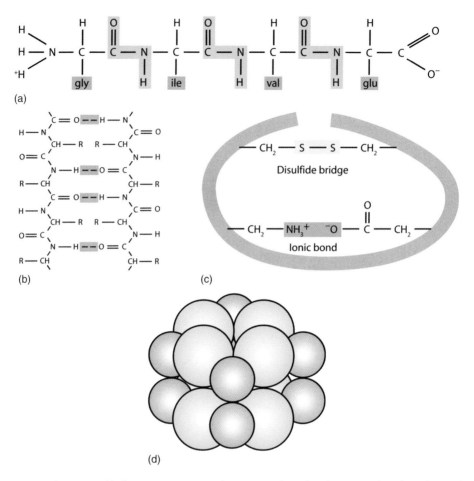

Figure 9.14 (a) The primary structure of a protein is the order of amino acids in the polypeptide chain. (b) Secondary structure results from hydrogen bonding. (c) The interaction between amino acids—due to the properties of their side chains—results in the tertiary structure of a protein, stabilized by disulfide bonds. (d) If a protein is composed of more than one protein chain, the assembly of the subunits into a protein complex constitutes its quaternary structure.

enzyme will fold into a three-dimensional shape that is dependent on its sequence of amino acids and the interaction of the different amino acids that make up the enzyme with one another. Like ions, some amino acids carry a positive charge and some carry a negative charge (Figure 9.12), and this also influences their folding and their interaction within the cell or the plasma membrane. Enzymes may also contain nonprotein cofactors that are necessary for their function, such as the iron (Fe) in the heme structure of the cytochromes in electron transport chains. Enzymes contain an active site into which the substrate or substrates of a reaction become bound. By orienting the substrate(s) correctly, the activation energy of the reaction is lowered, and a covalent bond is formed or broken.

Enzymes are catalysts for biological reactions

Chemical reactions usually either require an input of energy or result in a release of energy. Oxidation–reduction reactions release energy as electrons move between molecules. A **catalyst** is an agent that promotes change without itself being changed, and enzymes are protein catalysts. If we think of **biochemical pathways** as assembly lines, the enzyme catalysts are workers that perform a single step over and over on the product of the previous worker. The material they act on is their "substrate," and they produce a product that becomes the substrate of the next enzyme. Enzymes can be effective in small quantities because they are not used up in the reaction. They are extremely specific, so a given cell can contain several thousand different enzymes, each with its own job. Biochemical reactions can occur without enzymes, but enzymes speed up chemical reactions to the level that is needed to sustain life (Figure 9.15). This is the reason the cytosol contains so much soluble protein.

Enzymes act by binding with their substrate (the compound they alter) at their active site and are usually named by adding "-ase" to the name of the substrate (amylase works on amylose, sucrase works on sucrose, etc.) but they can also have a trivial name such as "rubisco," which is a nickname

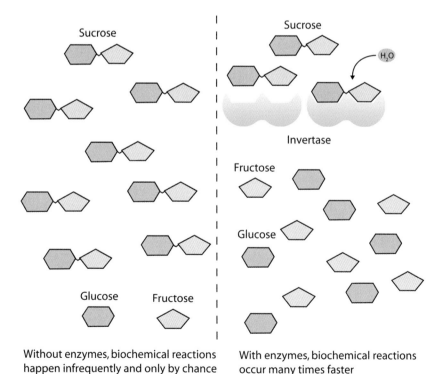

Figure 9.15 An enzyme is a catalyst that interacts with the substrates of a biochemical reaction and greatly increases the rate of a reaction.

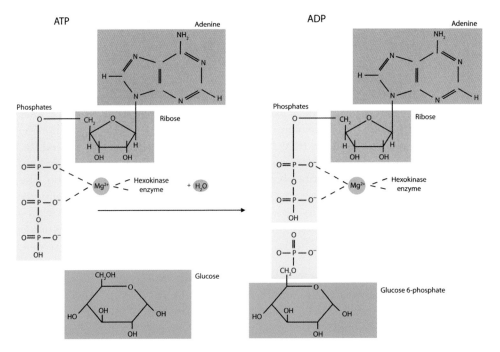

Figure 9.16 Some biochemical reactions are coupled with the transfer of a phosphate from ATP to increase the energy of the product in preparation for the next step in the pathway of the reaction. Here, the energy of glucose is increased by the addition of a phosphate bond from ATP in preparation for a successive reaction.

for the enzyme ribulose-1,5-bisphosphate carboxylase/oxygenase. This is the enzyme that adds CO_2 to the sugar ribulose-1,5-bisphosphate, which is the initial step in making new sugars as part of photosynthesis. The enzyme rubisco is present in large quantities in every photosynthetic cell, so there is more of this protein on earth than any other.

Some biochemical reactions cannot occur with the help of an enzyme alone because the reaction is energetically unfavorable. An analogy is that ice will melt at room temperature, but energy must be added to form ice in a room temperature environment. Energetically uphill biochemical reactions can occur, however, if they are coupled with the release of energy from ATP. One example is the synthesis of sucrose from a molecule of glucose and fructose by sucrose synthase, the reverse of the reaction is shown in Figure 9.15. This reaction requires the input of two molecules of ATP for each molecule of sucrose synthesized. In other cases, the energy of a phosphate bond is transferred from ATP to another molecule such as glucose (Figure 9.16). This phosphorylation energizes the recipient molecule and is used in subsequent reactions.

The rate of enzyme reactions can be controlled through the rate of enzyme synthesis and breakdown, or by activators and inhibitors (Figure 9.17). These may bind at the active site; in the case of competitive inhibitors, they

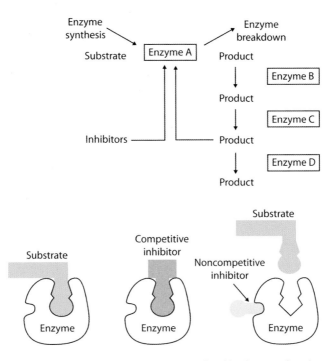

Figure 9.17 The rate of an enzyme reaction can be regulated by the rate of synthesis or breakdown of the enzyme. It may also be altered by the buildup of a product (feedback inhibition) or by the intervention of inhibitor molecules. Inhibitors may simply occupy the enzyme active site, or alter the conformation of the active site by binding to the enzyme at another point. (Redrawn and modified from Stryer 1975.)

may act through product or end-product inhibition, or occupy another site on the enzyme that creates, enhances, or reduces the activity of the enzyme by altering the active site. Noncompetitive inhibitors may change the active site so the substrate cannot bind or cannot be released.

Chapter 10

Photosynthesis

What is **photosynthesis**? It is the metabolic process by which plants absorb solar radiation in the range of visible light and convert this light energy into chemical energy. There are steps called the **photochemical (light) reactions of photosynthesis** (Figure 10.1) that produce NADPH (an electron-acceptor molecule) and ATP (energy held in phosphate bonds), and further steps called the **biochemical (dark) reactions** (Figure 10.1) that utilize this NADPH and ATP to convert carbon dioxide (CO_2) and water (H_2O) into sugars ($C_6H_{12}O_6$); these sugars have the general formula CH_2O. These sugars are used for respiration or growth in the leaves where they are produced, they are transported for use in other parts of the plant, or they are stored for future use. These sugars form the basis of the food chain for all life.

Light and photosynthesis

Within the range of wavelengths of all electromagnetic energy, the visible light used by photosynthesis is bracketed by shorter wavelength ultraviolet (UV) and longer wavelength infrared (IR) light. UV radiation has such high energy that it damages biological molecules by knocking electrons off, thus UV radiation is called "ionizing radiation." UV radiation is absorbed by glass, which is why you cannot get sunburn working in a greenhouse. IR radiation has too little energy to be useful in photosynthesis. The energy of IR radiation is absorbed by cells and produces heat. IR radiation is not absorbed by glass but passes through, so a greenhouse or car will become heated by sunlight. Visible light has just the right amount of energy for photosynthesis: enough to increase the energy level of electrons of the pigments that absorb light in the visible range but not enough to harm these molecules.

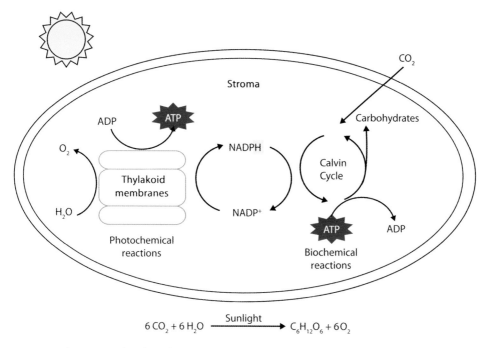

Figure 10.1 The photochemical (light) reactions of photosynthesis, on the left side of the diagram, take place on the thylakoid membranes of the chloroplast and the biochemical (dark) reactions, on the right side of the diagram, take place in the stroma of the chloroplast. The ATP and NADPH created by the photochemical reactions are used in the biochemical reactions.

Chloroplasts

Photosynthesis occurs in **chloroplasts** (Figure 1.14). These are football-shaped organelles consisting of an outer and an inner membrane that enclose **thylakoid membranes** in a gel-like inner matrix called the **stroma**. The thylakoid membranes are long, flattened membranous tubes where the absorption of light and production of chemical energy occur. In some parts of the stroma, the thylakoids are folded and stacked to form **grana**. The stroma is where the production of carbohydrates occurs, using the energy harvested from light by photosynthesis.

Like other membranes, the thylakoid membranes are made of lipids, although these lipids contain the sugar galactose (making them galactolipids) or sulfate (making a sulfolipid) as the head group, instead of phosphate as in phospholipids. The use of galactolipids in photosynthetic membranes reduces very significantly the amount of phosphorus required by plants and allows photosynthesis to continue under low phosphorus availability. The photosynthetic pigments are all hydrophobic to some degree, and therefore are embedded in the lipid bilayer of the thylakoid membranes.

Photosynthetic pigments

The **action spectrum for photosynthesis** (Figure 10.2a), or the range of light wavelengths that result in significant photosynthesis, is an indication of the sum of the absorption spectra of the pigments involved in photosynthesis.

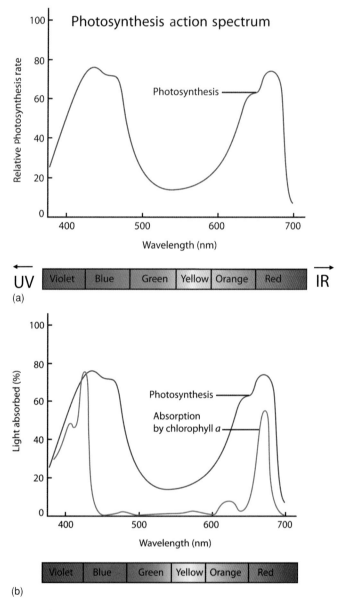

Figure 10.2 The action spectrum of photosynthesis (a) shows the relative levels of photosynthesis that occur over the visible light spectrum. The cumulative absorption of the photosynthesis pigments chlorophyll *a* (b), chlorophyll *b* (c), and the carotenoids (d) follows the general outline of the action spectrum.

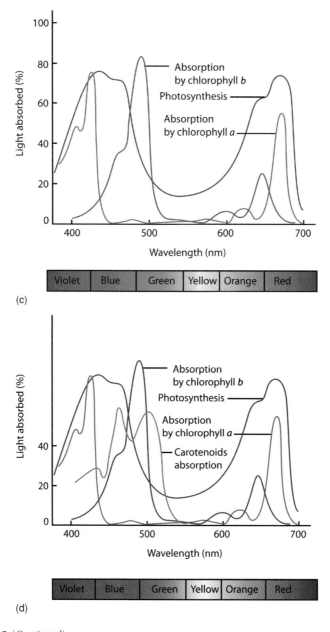

Figure 10.2 (*Continued*)

Chlorophyll

Chlorophyll *a* (chl *a*) is the primary pigment for light absorption in photosynthesis (Figure 10.3). It consists of a hydrophobic tail that is imbedded in the thylakoid membranes of the chloroplast, and a tetrapyrrole ring composed of four pyrrole rings, each with four carbon atoms and one nitrogen

162 Structure and Function of Plants

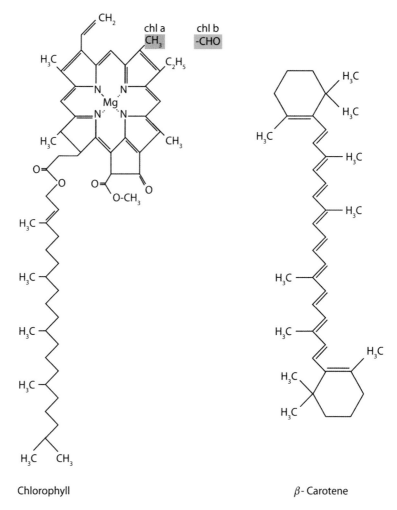

Figure 10.3 The photosynthetic pigments chlorophyll a and b (left) and β-carotene. Chlorophyll has a hydrophobic tail and the entire β-carotene molecule is hydrophobic. These hydrophobic regions are embedded in the thylakoid membrane lipid bilayer.

atom. Together, the four pyrrole rings sequester (hold on to) a magnesium atom. Chlorophyll a has two peaks of light absorption, one at a wavelength of 430 nm, in the violet-blue range, and the other at a wavelength of 662 nm, in the red range (Figure 10.2b). The action spectrum for photosynthesis does not precisely match the absorption spectrum of chlorophyll *a*, which suggests the involvement of other pigments.

Accessory pigments

Chlorophyll *b* (chl *b*), like chl *a*, has two peaks of light absorption, one at 453 nm and the other at 642 nm, which is more orange than red (Figure 10.2c). Plants usually contain about half as much chl *b* as chl *a*. The other important accessory pigments are the **carotenoids**, which are red, orange, and yellow in color. The maximum absorption of the carotenoids is between 460 and 550 nm, in the blue and green range (Figure 10.2d). The carotenoids are entirely hydrophobic (Figure 10.3), and are therefore fully embedded in the membrane. When the absorption spectra for chl *a* and the accessory pigments involved in photosynthesis are added together, they approximate the action spectrum for photosynthesis.

The "light" reactions of photosynthesis

Light energy is captured by **antenna complexes** (Figure 10.4a), groups of chloroplast pigments arranged in aggregates on thylakoid membranes, and consists of about 300 molecules of chl *a* and 50 molecules of carotenoids and chl *b*. The energy flows to a chl *a* molecule at the reaction center of the complex.

It was discovered in the early 1950s that while chlorophyll absorbs light at wavelengths longer than 690 nm, light in this range alone was not sufficient for photosynthesis. It was also shown that when light of wavelengths longer than 690 nm was supplemented by visible light of shorter wavelengths, the photosynthetic rate was greater than that produced by the light of shorter wavelengths alone. This is known as the **Emerson enhancement effect**, named for Robert Emerson, the scientist who made the discovery.

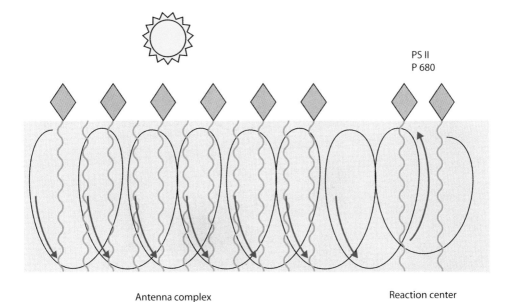

(a)

Figure 10.4 The photochemical or "light" reactions of photosynthesis: (a) An antenna complex is a cluster of a few hundred molecules of chl *a*, chl *b*, and other pigments including carotenoids. Light energy accumulated by these accessory pigments in the antenna complex is funneled to the reaction center, which consists of a molecule of chl *a* and a receptor molecule.

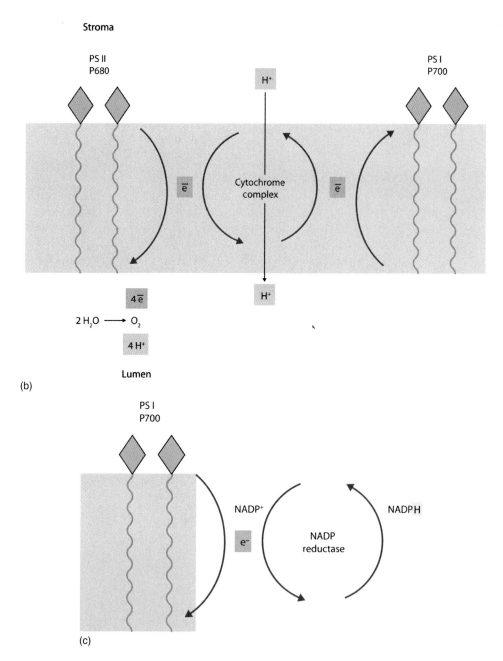

Figure 10.4 (*Continued*) (b) Electrons move from chlorophyll P680 in Photosystem II (PS II) through a series of electron carriers (the electron transport chain—Figure 10.5) that are successively reduced and oxidized as they pass the electrons along. Oxygen (O_2) is generated when electrons from water (H_2O) are used to replace those lost by P680. During the movement of electrons along the electron transport chain, protons (H^+) are pumped from the stroma into the lumen of the thylakoid membranes. (c) At Photosystem I (PS I), electrons are transferred to the P700 chl *a* molecule. Light energy absorbed by the Photosystem I antenna complex reenergizes the electrons and they pass through another electron transport chain (Figure 10.5). The last step in the photosynthetic process is the reduction of $NADP^+$ by two electrons to form NADPH.

The reason for the Emerson enhancement effect is that there are two distinct **reaction center** chlorophylls, one with a chl *a* that absorbs maximally at 700 nm called P (for "pigment") 700, and the other with a chl *a* that absorbs maximally at 680 nm, called P680. The complex containing **P700** is called Photosystem I, and the complex containing **P680** is called Photosystem II. These photosystems were named in order of their discovery, but in the process of photosynthesis, Photosystem II precedes Photosystem I (Figure 10.4b).

The photosystems contain pigments, proteins, other compounds, and several different ions, including manganese, iron, calcium, and chlorine. The function of the photosystems is to absorb light energy and pass electrons excited by this light energy absorption into a chain of associated molecules located in or on the thylakoid membranes which is called an **electron transport chain** (Figure 10.5). As the electrons move through the electron transport chain, they lose energy, which is used to pump hydrogen ions (protons) into the lumen of the thylakoid membrane (Figure 10.4b). These protons are used to create ATP from ADP and inorganic phosphate (**photophosphorylation**), and the electrons are used to reduce $NADP^+$ to NADPH (Figure 10.4c). Both ATP and NADPH are required for the production of sugars in the subsequent biochemical reactions of photosynthesis.

Electrons to replace those released at the reaction center come from water, which is split to release its components: two electrons, two hydrogen ions (protons or H^+), and an atom of oxygen, which combines with another

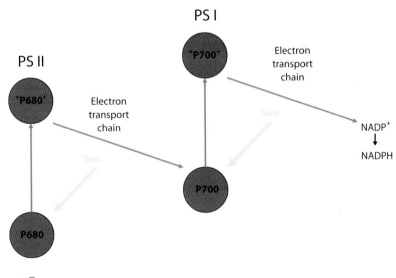

Figure 10.5 The energy of light is captured by an antenna complex and transferred to P680 in Photosystem II. This increased energy (*P680*) causes electrons to move from P680 through a series of electron carriers until they reduce P700 in Photosystem I. P680 was oxidized by the loss of electrons which are replaced with electrons from water, generating protons and oxygen. P700 receives light energy (*P700*) from its own antenna complex, and the reenergized electrons pass through another series of electron carriers, finally reducing $NADP^+$ to NADPH.

oxygen atom from the same process to produce the gas molecular oxygen (O_2), which is released from the plant as a by-product of photosynthesis (Figure 10.4b). This is the reaction that causes plants to be net producers of oxygen for the biosphere.

pH utilizes a log scale, so a decrease of one pH unit represents a 10-fold increase in the number of hydrogen ions (Chapter 7). A decrease from pH 8 to pH 5 is a change of 3 pH units or a 1000-fold increase in the number of H^+. Low pH (e.g., 5) indicates acidity or a high concentration of protons (H^+), while a high pH (e.g., pH 8) indicates alkalinity or a low concentration of protons. In the light, the pH of the stroma will rise to about 8, while the pH of the thylakoid lumen (the space inside the tubular thylakoid membranes) will decrease to about 5, which means that there are about 1000 times more hydrogen ions (H^+) in the thylakoid lumen than in the stroma.

The hydrogen ions that buildup in the thylakoid lumen, creating a low pH, are able to flow back into the stroma through a protein complex called **ATP synthase**, which harnesses the energy generated by the flow of hydrogen ions to synthesize ATP by adding an inorganic phosphate group ($H_2PO_4^-$) to ADP (Figure 10.6). During the light reactions of photosynthesis, both ATP and NADPH accumulate in the stroma, where they can be used in the synthesis of carbohydrates. These reactions are called the "light" or photochemical reactions of photosynthesis because they utilize light energy to create the chemical energy of ATP and NADPH, and therefore only take place in the light.

The "dark" reactions of photosynthesis

As long ago as 1905, it was determined that in dim light, altering the temperature at which photosynthesis was occurring did not affect the rate of photosynthesis, but increasing the light intensity did. However, in bright light, the rate of photosynthesis was affected by change in temperature, while a further increase in light had no effect. The rate of biochemical reactions of living organisms increases with increasing temperature within the temperature range at which the organism thrives. Investigation of these differences led to the early understanding of photosynthesis as a group of photochemical ("light") reactions linked through ATP and NADPH to a group of biochemical ("dark") reactions. In dim light, photosynthetic rate is limited by the photochemical (or light) reactions in which the energy of light is captured and converted to ATP and NADPH. However, in bright light, the rate of the light reactions is high, but photosynthesis can be limited by the rate of the associated biochemical (or dark) reactions that use the energy created in the light reactions to make carbohydrates.

In reality, both the light and dark reactions of photosynthesis occur only in the light. The biochemical reactions are termed the dark reactions simply to indicate that light intensity is not a direct regulatory factor. In the dark, when the flow of chemical energy from the light reactions has stopped, then the dark reactions stop for lack of ATP and NADPH.

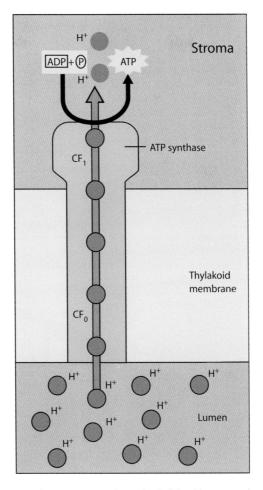

Figure 10.6 The protons that were pumped into the thylakoid lumen as electrons moved through the electron transport chain from Photosystem II to Photosystem I flow from the lumen back into the stroma through a protein complex called ATP synthase. ATP synthase is composed of CF_0, the membrane-bound channel through which protons flow, and CF_1 where ATP is formed. ATP synthase uses the energy generated by the movement of these H^+ to synthesize ATP from ADP and inorganic phosphate (P). (Diagram adapted from Wilkins 1988, used with permission.)

The sum of the light and dark reactions of photosynthesis creates carbohydrates (CH_2O) from carbon dioxide (CO_2) and water (H_2O), with oxygen (O_2) as a by-product. Determination of the pathway for the biochemical reactions of photosynthesis was done by identifying the compounds and the sequence of their synthesis in photosynthesis. The first product of photosynthesis was determined to be a 3-carbon compound called **3-phosphoglyceric acid** or **3-PGA**. Therefore, the simplest form of photosynthesis came to be known as **C_3 photosynthesis**. The variations on C_3 photosynthesis used by some plants in stressful environments involve additional pathways and have been given other names, but they also still utilize the same basic reactions found in C_3 photosynthesis.

A Nobel Prize for the Biochemistry of Photosynthesis

Melvin Calvin won the Nobel Prize in Chemistry in 1961 for his discoveries related to the biochemistry of photosynthesis. Calvin earned a bachelor's degree in chemistry in 1931 from the Michigan College of Mining and Technology and a doctorate in 1935 from the University of Minnesota. He did postdoctoral work at the University of Manchester in England funded by a Rockefeller Foundation grant and was invited to join the faculty of the University of California at Berkeley in 1937. He worked on the purification of plutonium as part of the Manhattan Project during World War II, and began work on photosynthesis using the radioactive isotope of carbon ^{14}C in 1946, the same year he became director of the bioorganic chemistry group at the school's Lawrence Radiation Laboratory, now the Lawrence Livermore National Laboratory.

The work on the biochemical reactions of photosynthesis elucidated the sequence of steps in the pathway that used the products of the light reactions of photosynthesis to produce sugars. The ^{14}C radioactive isotope of carbon, discovered in 1940, made it possible to trace the sequence of these reactions. A solution of sodium bicarbonate ($NaHCO_3$) containing trace amounts of ^{14}C was fed to the algae *Chlorella pyrenoidosa* in lollipop-shaped flattened glass vessels (Figure 10.7). The algal cells

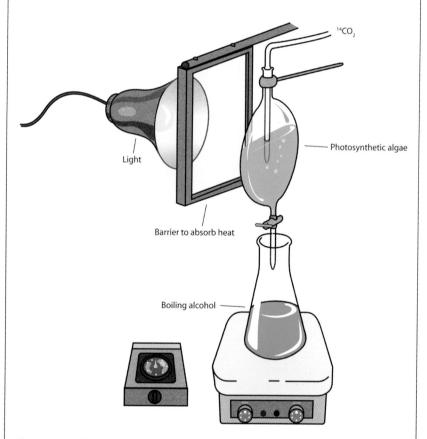

Figure 10.7 In dissecting the biochemical pathway of photosynthesis, Calvin exposed algal cells in flattened "lollipop" flasks to radiolabeled carbon ($^{14}CO_2$), illuminated them for short time periods, then killed the cells and extracted the newly synthesized compounds.

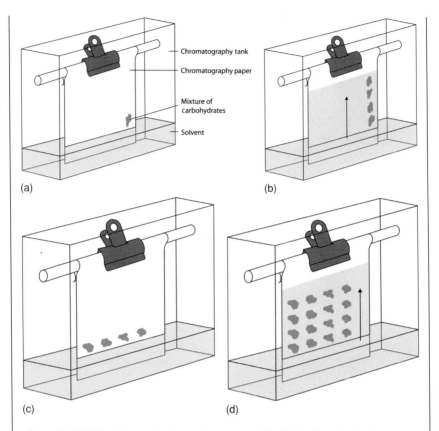

Figure 10.8 (a) The photosynthetic products extracted by Calvin from algal cells were concentrated and blotted onto a corner of chromatography paper which was suspended in a tank with a solvent that was absorbed by the chromatography paper. (b) As the solvent face moves along the paper, the compounds that are most soluble move the most quickly, separating them from slower-moving compounds. (c) The chromatography paper is dried overnight, turned 90°, and the edge suspended in a new solvent. (d) The compounds that overlapped after the first solvent run are separated into a second dimension, where they can be quantified and identified.

were illuminated for short periods of time and then the cells were killed by releasing them into alcohol. The organic compounds synthesized during illumination would be labeled with ^{14}C and would therefore be radioactive.

The organic compounds were extracted from the algae, concentrated, and then individual compounds were separated using paper chromatography. A small amount of the extract was blotted onto a corner of chromatography paper and the paper suspended with one edge in a solvent (Figure 10.8a). As the solvent face traveled across the paper, the most soluble compounds moved most quickly, separating the mixture of compounds along one side of the paper (Figure 10.8b). However, when separated using a single solvent, some compounds overlapped with others. The paper was removed from the chromatography tank, allowed to dry, and turned 90°, so the separated spots were closest to the solvent reservoir (Figure 10.8c). A different solvent was used to separate the existing spots in a second dimension as the solvent face moved along the paper (Figure 10.8d).

To visualize the compounds, the chromatography paper was placed in contact with X-ray film. The compounds of interest would have incorporated ^{14}C and would therefore expose the X-ray film, showing the location of each compound on the paper. The quantity of each compound was determined by exploring the chromatography paper with a Geiger counter to determine the level of radioactivity, and compounds were analyzed biochemically to determine their identity.

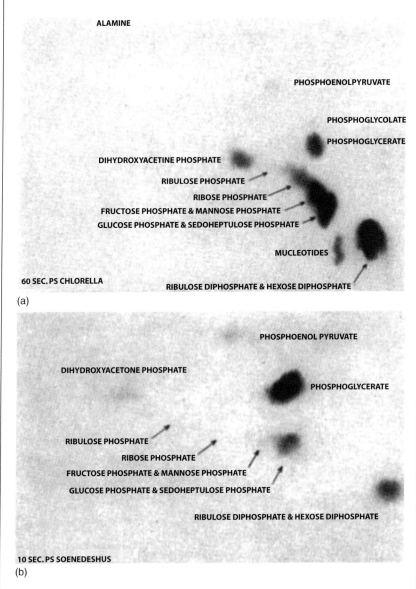

Figure 10.9 (a) The chromatography paper is dried and put in contact with X-ray film. Because ^{14}C is radioactive, it exposes the film, creating autoradiographs like the one seen here. When the algae were killed and extracted after 60 seconds of photosynthesis, there were several newly synthesized compounds. (b) When algae were illuminated for only 10 seconds, one compound, phosphoglycerate (3-PGA), contained most of the radioactivity. This provided strong evidence that 3-PGA was the first product of the biochemical reactions of photosynthesis. (Reprinted with permission from Energy Reception and Transfer in Photosynthesis; Calvin 1959; copyright 1959 by the American Physical Society.)

> From the autoradiographs, it can be seen that when the algae were exposed to light in the presence of ^{14}C for 60 seconds a number of compounds were produced (Figure 10.9a). To determine the earliest product of photosynthesis—the first step in the synthesis of sugars—the length of exposure to light was shortened. When the experiment was run for only 10 seconds, a single compound, 3-phosphoglyceric acid, was found in the largest concentration (Figure 10.9b). The photosynthetic carbon pathway was elucidated by 1957, and is now called the Calvin cycle in honor of Melvin Calvin (Calvin 1959, Encyclopædia Britannica online, February 6, 2008).

The reactions that produce 3-PGA add carbon dioxide to a 5-carbon sugar, ribulose bisphosphate (RuBP), producing a 6-carbon molecule that immediately splits into two molecules of 3-PGA (Figure 10.10). The enzyme that catalyzes this reaction is **ribulose bisphosphate carboxylase/oxygenase**, and it has been nicknamed rubisco (RuBisCO). This enzyme is the most abundant protein on the earth. The series of reactions that produce 3-PGA and that eventually regenerate RuBP are called the **Calvin cycle**, named for Melvin Calvin, who led the work that elucidated this pathway.

As 3-PGA accumulates during the biochemical reactions of photosynthesis, the available phosphate in the stroma becomes tied up, so 3-PGA is used to make glucose, releasing inorganic phosphate, and the glucose is used to make starch because starch is not osmotically active (it will not attract water). At the end of the day, large grains of starch can be seen in the stroma of chloroplasts. At night, the starch is broken down again to 3-carbon carbohydrates (Figure 11.4). In the cytosol, two of these 3-carbon sugars are used to make either glucose or fructose, and one of each of these molecules is linked together to make one molecule of sucrose, the sugar that is used for translocation (movement of organic molecules) from one part of the plant to another.

Photorespiration

When plants are under drought stress, there may be sufficient light for photosynthesis, but the stomata will close to prevent further water loss. Ongoing photosynthesis will utilize the available CO_2, resulting in a low concentration of CO_2 molecules for the biochemical reactions of photosynthesis inside the leaf. The "active site" of the enzyme rubisco, where CO_2 is added to the RuBP sugar molecule will also accept an oxygen (O_2) molecule instead of CO_2. This is why rubisco has both a "c" and an "o" at the end of its name: it is both a carboxylase and an oxygenase enzyme.

When the concentration of CO_2 is low, there is still plenty of O_2 available, because the concentration of CO_2 in the atmosphere is about 380 ppm or 0.034%, while the concentration of O_2 is about 21% or 210,000 ppm in the atmosphere. For other factors that affect the activity of rubisco as an oxygenase, see the box on page 173.

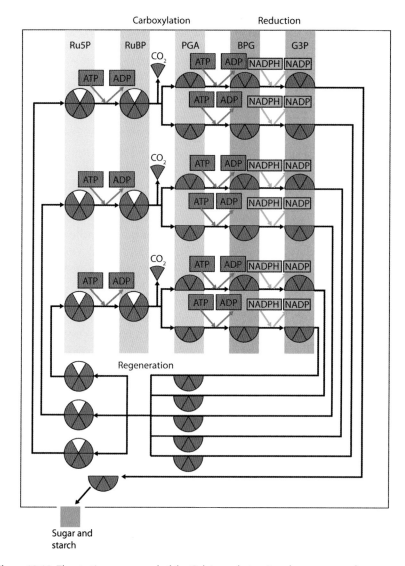

Figure 10.10 The starting compound of the Calvin cycle is a 5-carbon compound ribulose-5-phosphate (Ru5P). It is energized by the addition of a second phosphate bond to form ribulose-1,5-bisphosphate (RuBP; "bis" is a variation on "bi," meaning two). In the carboxylation step, carried out by the enzyme rubisco (ribulose-1,5-bisphosphate carboxylase/oxygenase), one carbon (C) as CO_2 is added to RuBP to make a 6-carbon compound that quickly separates into two molecules of 3-PGA. In successive steps of the cycle, a phosphate bond is transferred from ATP to form 1,3-bisphosphoglycerate (BPG), and in the reduction step, NADPH transfers electrons to BPG to form glyceraldehyde-3-phosphate (G3P). These are the reactions that use the products of the photochemical reactions of photosynthesis. In regeneration, many steps are required to form the starting compound of the cycle, RuBP. The diagram shows the net gain of one G3P (3 C) for every three carboxylations. (Diagram adapted from Wilkins 1988, used with permission.)

The enzyme rubisco can add either CO_2 or O_2 to RuBP because they both fit into rubisco's active site. What factors influence the balance between photosynthesis and photorespiration?

1. The concentration of O_2 in ambient air is nearly 21%, while the concentration of CO_2 in air is only 0.038% or 380 ppm (parts per million), and increasing a few ppm each year. The ratio of molecules of CO_2 to O_2 in air is therefore about 0.00164.
2. Cells in the leaf are covered by a thin film of water, so both CO_2 and O_2 become dissolved before they reach the rubisco enzyme. The solution concentration of CO_2 in water is relatively high compared to that of O_2. Therefore, at room temperature (25°C or 77°F), the ratio of dissolved CO_2 to O_2 is 0.0416 (Taiz and Zeiger 2006).
3. The affinity of rubisco for CO_2 is as much as 80 times greater than the affinity of rubisco for O_2. Therefore, rubisco will generally add CO_2 to RuBP in preference to O_2. However, the affinity of rubisco for O_2 *increases* with temperature, and the ratio of CO_2 to O_2 in solution *decreases* with temperature.
4. Under hot, dry conditions, such as drought, stomata will close to reduce transpiration but photosynthesis will continue, drawing down the internal concentration of CO_2 and thus decreasing the ratio of CO_2 to O_2 and increasing the incidence of photorespiration.

The coupling of O_2 to RuBP is called **photorespiration,** because it results in the loss of CO_2 (Figure 10.11), and is a significant inefficiency of the C_3 photosynthetic process. Plants are the source of atmospheric O_2, so in

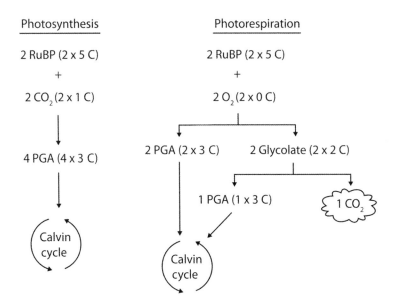

Figure 10.11 Rubisco will react with CO_2 in preference to O_2, but when the internal concentration of CO_2 in leaves is reduced by stomatal closure in light, the result is photorespiration. The oxygenase function of rubisco allows it to add O_2 to RuBP. The products are one molecule of 3-phosphoglycerate (PGA) and one molecule of glycolate. For every two molecules of glycolate produced, one CO_2 is lost to the atmosphere, so photorespiration costs the plant carbon while photosynthesis produces a net carbon gain.

the early evolution of plants, O_2 concentration in the air was very low. Photorespiration was not a problem until O_2 concentrations increased. In photorespiration, because no carbon atom has been added to RuBP, the product of RuBP plus O_2 still has only five carbons. One 3-PGA is synthesized rather than two, and a molecule of glycolate is made from the other two carbons. Two molecules of glycolate can be used to make one additional 3-PGA, but one CO_2 is lost (thus the name photorespiration). Temperate plants at low internal concentrations of CO_2 make no net sugar through photorespiration and lose a carbon for every two molecules of O_2 that is added to RuBP.

Adaptations to make photosynthesis more efficient

C_4 photosynthesis

A number of plants utilize a variation of C_3 photosynthesis (called "C_4 photosynthesis") that is more efficient under hot, dry conditions, such as conditions during the summer on the Great Plains. C_4 photosynthesis involves locating the C_3 Calvin cycle reactions of photosynthesis in a group of cells that cluster around the veins in the leaves of these plants, many of which are grasses. These cells are called the bundle sheath cells, and they are surrounded and protected from exposure to the atmosphere by mesophyll cells (Figure 10.12a). In the leaves of C_3 grasses, veins also are surrounded by bundle sheath cells, but they usually do not have chloroplasts and function

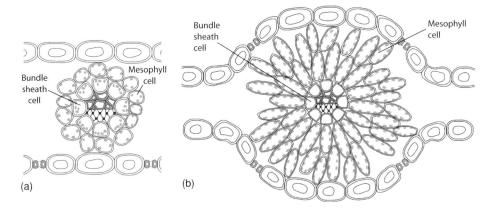

Figure 10.12 In the leaves of both C_4 and C_3 grasses, the xylem and phloem of veins are surrounded by bundle sheath cells, which in turn are surrounded by mesophyll cells. (a) In C_4 leaves, CO_2 is captured in mesophyll cells and shuttled into bundle sheath cells, where it is released for the Calvin cycle of photosynthesis. Because the CO_2 concentration at rubisco can be kept high in C_4 plants by this shuttle, photorespiration is essentially eliminated even under the hot, dry conditions that cause stomata to close. (b) In C_3 leaves, Calvin cycle photosynthesis takes place in mesophyll cells. If stomata close and CO_2 concentrations decrease, the rubisco in these leaves will react with O_2 and photorespiration will occur.

to load sucrose into the phloem (Figure 10.12b). Mesophyll cells surround the bundle sheath in these leaves as well, but their chloroplasts carry out C_3 photosynthesis. In C_4 photosynthesis, CO_2 is captured in the mesophyll cells and used to synthesize the 4-carbon compound oxaloacetate (OAA). Because OAA is a 4-carbon compound, the name C_4 photosynthesis is given to this system. Maize is an important crop plant that utilizes C_4 photosynthesis.

The enzyme that catalyzes the capture of CO_2 in this system, **phosphoenolpyruvate carboxylase** (PEP carboxylase or **PEPcase**), unlike rubisco, has no activity toward oxygen. OAA is converted to other 4-carbon organic acids that move from the mesophyll into the bundle sheath cells, dropping off CO_2 in the bundle sheath cells where it can be used for the Calvin cycle or C_3 photosynthesis (Figure 10.13). Because the concentration of CO_2 is

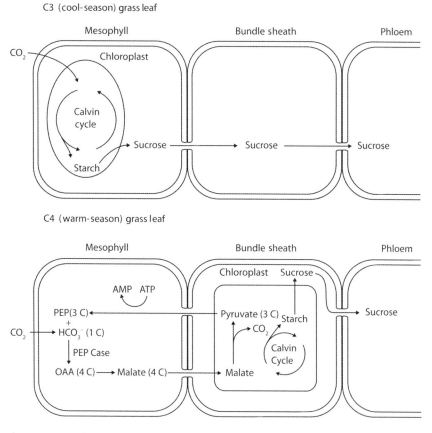

Figure 10.13 In the mesophyll cells of C_4 plants, CO_2 is added to phosphoenolpyruvate (PEP) by the enzyme PEP carboxylase, which does not react with O_2. PEP is a 3-carbon compound, so the first product of photosynthesis is a 4-carbon compound. The 4-carbon compound that carries CO_2 from mesophyll to bundle sheath cells is malate or aspartate depending on the plant species. In the bundle sheath, CO_2 is released for use in the Calvin cycle and the remaining 3-carbon compound, pyruvate, is returned to the mesophyll. However, formation of PEP from pyruvate requires the energy of two phosphate bonds from ATP. This energy cost makes C_4 photosynthesis more efficient than C_3 photosynthesis only for plants in hot, dry climates that would otherwise be subject to photorespiration.

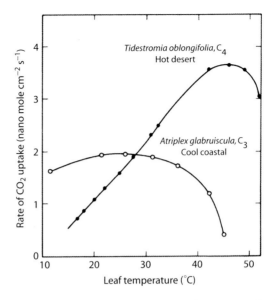

Figure 10.14 Photosynthesis as a response to temperature of a C_3 species (*Atriplex glabriuscula*) native to and grown in a cool coastal (16°C/10°C day/night) environment and a C_4 species (*Tidestromia oblongifolia*) native to and growing actively in the hottest part of the year in Death Valley (45°C/31°C day/night). Rate of CO_2 uptake for the C_3 plant is high to about 30°C (86°F), but declines with increasing temperature. The hot desert-adapted plant reached its maximum photosynthetic rate at about 45°C (113°F). C_3 plants are most efficient at lower temperatures, while C_4 plants are more efficient under hot dry conditions when internal CO_2 concentrations can be low. (Figure from Bjørkman et al. 1975, used with permission.)

always kept high in the bundle sheath cells of C_4 plants, no photorespiration occurs. However, because the energy of two phosphate bonds from one ATP (producing AMP or adenosine monophosphate) is required for the synthesis of the 3-carbon molecule PEP to which CO_2 becomes attached, the C_4 photosynthetic process is only more efficient under the hot, dry conditions that would cause photorespiration (Figure 10.14). There is evidence that the ancestors of humans evolved to be bipedal (or walk on two feet) at about the same time that C_4 photosynthesis evolved. In both cases, the change was driven by a hotter, drier climatic period.

CAM photosynthesis

Another variation of C_3 photosynthesis occurs in some desert succulents. In this scheme, plants open their stomata only at night (Figure 8.11), when transpiration is reduced and CO_2 can be absorbed from the atmosphere and used to form the organic acid malic acid, which is stored in the vacuole (Figure 10.15). During the day, when there is light available for photosynthesis, the CO_2 is released from malic acid (or malate, the charged form) and used

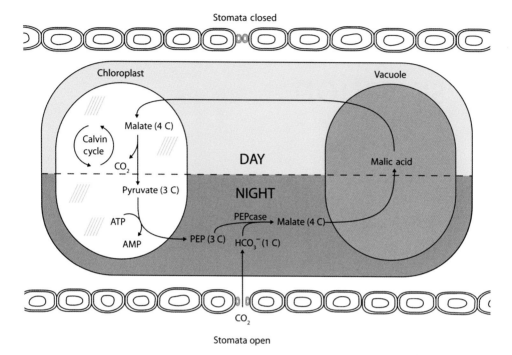

Figure 10.15 Photosynthesis of succulent CAM (Crassulacean acid metabolism) plants is uniquely adapted to dry desert environments. Stomata open at night when transpiration is minimal for the uptake of CO_2, which accumulates in the vacuole as malic acid. During the light period, when ATP and NADPH are available from the photochemical reactions of photosynthesis, malate releases CO_2 for the Calvin cycle. The biochemistry supporting the day/night or temporal sequestration of CO_2 capture and release in CAM plants resembles the spatial sequestration in mesophyll and bundle sheath cells in C_4 species.

in the Calvin cycle for C_3 photosynthesis, even though the stomata remain shut. The synthesis and utilization of malic acid result in significant changes in vacuolar pH from day to night. This variation of photosynthesis is called **Crassulacean acid metabolism** or CAM, because it was originally discovered in the Crassulacean plant family. Pineapple is a crop plant that utilizes CAM photosynthesis.

Chapter 11

Respiration

When viewed as a summary of the biochemical reactions it involves, respiration is the reverse of photosynthesis. In **respiration**, oxygen is used, carbohydrates are enzymatically broken down to CO_2, oxygen is reduced to water, and chemical energy (ATP) is synthesized. The energy released by respiration originated as solar energy and was transferred to chemical energy in photosynthesis where it was stored as carbohydrates. The energy tied up in sugars is liberated by respiration and used to form molecules with "reducing power" (NADH and $FADH_2$). The reducing power is used in the electron transport chain, the last steps of respiration, to make large quantities of ATP, the end product of respiration.

When and where does respiration occur?

Respiration occurs in essentially all the cells of plants and animals all the time, to supply the energy needed for the work done by enzymes. Over the life of a plant or a plant organ, respiration is greatest during initial growth, and there is another period of increased respiration during senescence (the removal of proteins and minerals from leaves in the fall) or during fruit ripening (Figure 11.1).

For plants to accumulate dry matter, the rate of fixation of CO_2 by photosynthesis must exceed the rate of CO_2 evolution from respiration. Both processes are affected by temperature, but respiration has a higher optimal temperature than photosynthesis. At temperatures higher than the optimal for photosynthesis, respiration continues to increase and the rate of photosynthesis falls, decreasing growth rate (Figure 11.2).

Plants with C_4 photosynthesis are better adapted to combinations of high temperature and drought than C_3 plants. While both C_3 and C_4 plants employ the Calvin cycle of photosynthesis, in C_4 plants, photosynthesis occurs in bundle sheath cells. CO_2 is captured by the surrounding mesophyll cells,

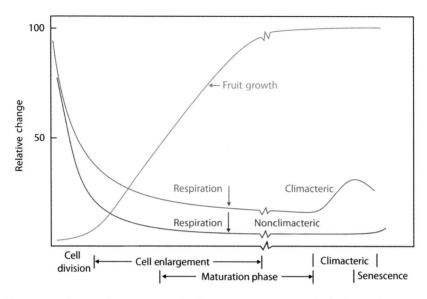

Figure 11.1 The rate of respiration is related to enzymatic activity and is highest in fruits during their early growth (green line), which occurs primarily by cell division. Respiration declines during cell enlargement as enzyme concentration is diluted by water uptake for cell growth. At the end of the maturation stage, there is a rise in respiratory activity termed the climacteric that is associated with fruit ripening (red line). In banana (*Musa acuminata*), most of the starch in the fruit is converted to sugar during the climacteric. Dry weight decreases measurably during the climacteric due to respiration. In other fruits, such as citrus, no climacteric rise in respiration is observed during fruit ripening (blue line). (Figure from Biale 1964, used with permission.)

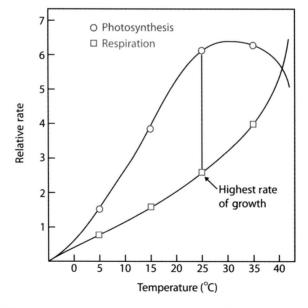

Figure 11.2 Effect of temperature on net photosynthesis (blue circles) and respiration (red squares) of field plants of the moss *Bryum sandbergii*. (Figure from Rastorfer and Higinbotham 1968, used with permission.)

and CO_2 concentration is kept high to prevent photorespiration, the unproductive loss of CO_2 during photosynthesis. C_4 plants are more efficient at higher temperatures because their stomata do not need to open for as long as that required by C_3 plants to supply the CO_2 needed for photosynthesis. Therefore, the optimum temperature required for photosynthesis is higher in C_4 than in C_3 plants (Figure 10.13).

Sources of energy for respiration

The first step in cellular respiration is the breakdown of the storage forms of carbohydrates, most commonly **starch** (as in the endosperm of grains), into glucose. Glucose and fructose from other storage forms of carbohydrates can also feed into cellular respiration. Sucrose, the form in which carbohydrates are transported in the phloem, is composed of one molecule of glucose and one molecule of fructose. **Fructans**, which are chains of fructose molecules with one glucose molecule at the end, are the storage forms of carbohydrates in onions and most small grains and temperate grasses. Fructans are composed of varying numbers of fructose molecules, from a few to as many as 200, and are stored in the vacuole because, unlike starch, they are water-soluble.

The sugars formed by photosynthesis accumulate in chloroplasts during the light period. 3-PGA, a 3-carbon sugar with a phosphate group attached (a triose phosphate) is used to form glucose, which is used to form starch (Figure 11.3). At night, the starch in chloroplasts is broken down to glucose and then to triose phosphates, which the chloroplast exchanges for inorganic phosphates (Pi). In the cytosol, the triose phosphates are used to make glucose and fructose, which in turn is used to make sucrose, the transport form of carbohydrates (Figure 11.4) and Pi is recycled. Sucrose is exported from mesophyll cells to bundle sheath cells and loaded into the phloem, as illustrated in Figure 5.24.

Cellular respiration

In all cells there is a constant need for energy to support the enzymatic activity required for growth and maintenance, and this is supplied as ATP from respiration. There are three stepwise processes involved in cellular respiration:

1. Glycolysis utilizes glucose or fructose and takes place in the cytosol of the cell. The end product of glycolysis is pyruvate plus a small amount of ATP and NADH.
2. The Krebs cycle takes place in the inner matrix of mitochondria and its main product is reducing power (NADH and $FADH_2$).
3. The last step of respiration, the electron transport chain, takes place on the **inner membrane** of mitochondria. This is the step where most ATP is synthesized and oxygen is used. The reducing power of NADH and

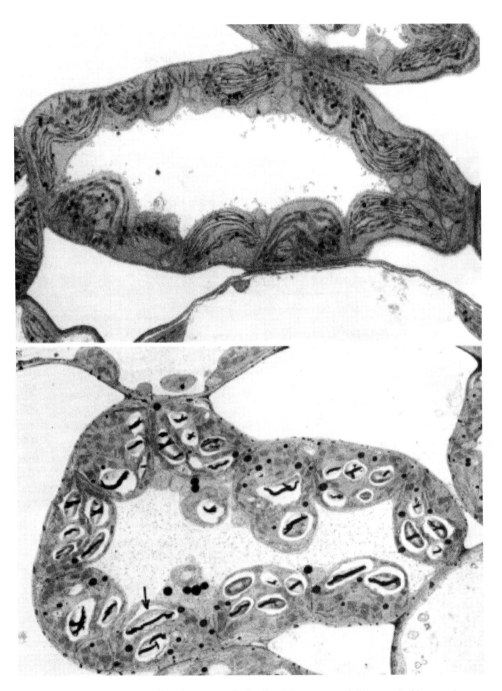

Figure 11.3 Chloroplasts from mesophyll cells of the youngest fully expanded leaves of soybean (*Glycine max*) grown in hydroponic culture at the full flower stage. Starch has accumulated in chloroplasts in the lower photo. In the upper photo, plants were supplied with optimal phosphate, while plants in the lower photo were given 1/10th this amount. With insufficient inorganic phosphate in the cytosol to exchange for triose phosphates from photosynthesis, starch accumulation in chloroplasts is excessive. (Figure from Lauer *et al.* 1989, used with permission.)

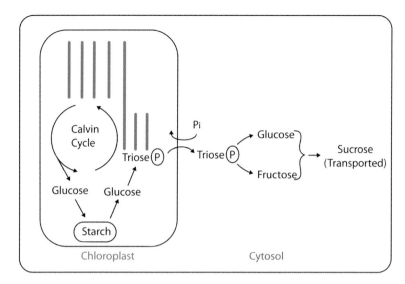

Figure 11.4 During the light period, carbohydrates from photosynthesis are used to make starch in chloroplasts. During the dark period, starch is broken down to triose (3-carbon) phosphates and exchanged for inorganic phosphates (Pi) from the cytosol. In the cytosol, sucrose is synthesized and exported.

$FADH_2$ is used by way of the electron transport chain to add a phosphate ($H_2PO_4^-$) group to ADP (adenosine diphosphate) to form **ATP (adenosine triphosphate)** (Figure 11.5).

The biochemistry of respiration is very similar in both plants and animals, and the pathways of respiration are linked to many other biochemical pathways. Intermediates of respiration are used to form amino acids, lipids, hormones, pigments, and nucleotides (Figure 11.6). This integration of respiration with other cellular processes means that much of the carbohydrate that enters respiration is diverted to the synthesis of products other than ATP.

Glycolysis

Glycolysis is the first step of respiration. It takes place in the cytosol and is fed by carbohydrates such as sucrose, starch, and fructan (Figure 11.7). Either glucose or fructose can be used as the initial substrate. In the first step, two ATPs are added to one molecule of glucose, and in the second step, phosphate is added to 6-carbon sugars. The name "glycolysis" refers to the lysis or splitting of sugars, and these hexose phosphates are then split into two triose phosphates, glyceraldehyde-3-phosphate, and dihydroxyacetone phosphate, the same compounds seen as intermediates of photosynthesis. These two triose phosphates can easily be interconverted. At this step of

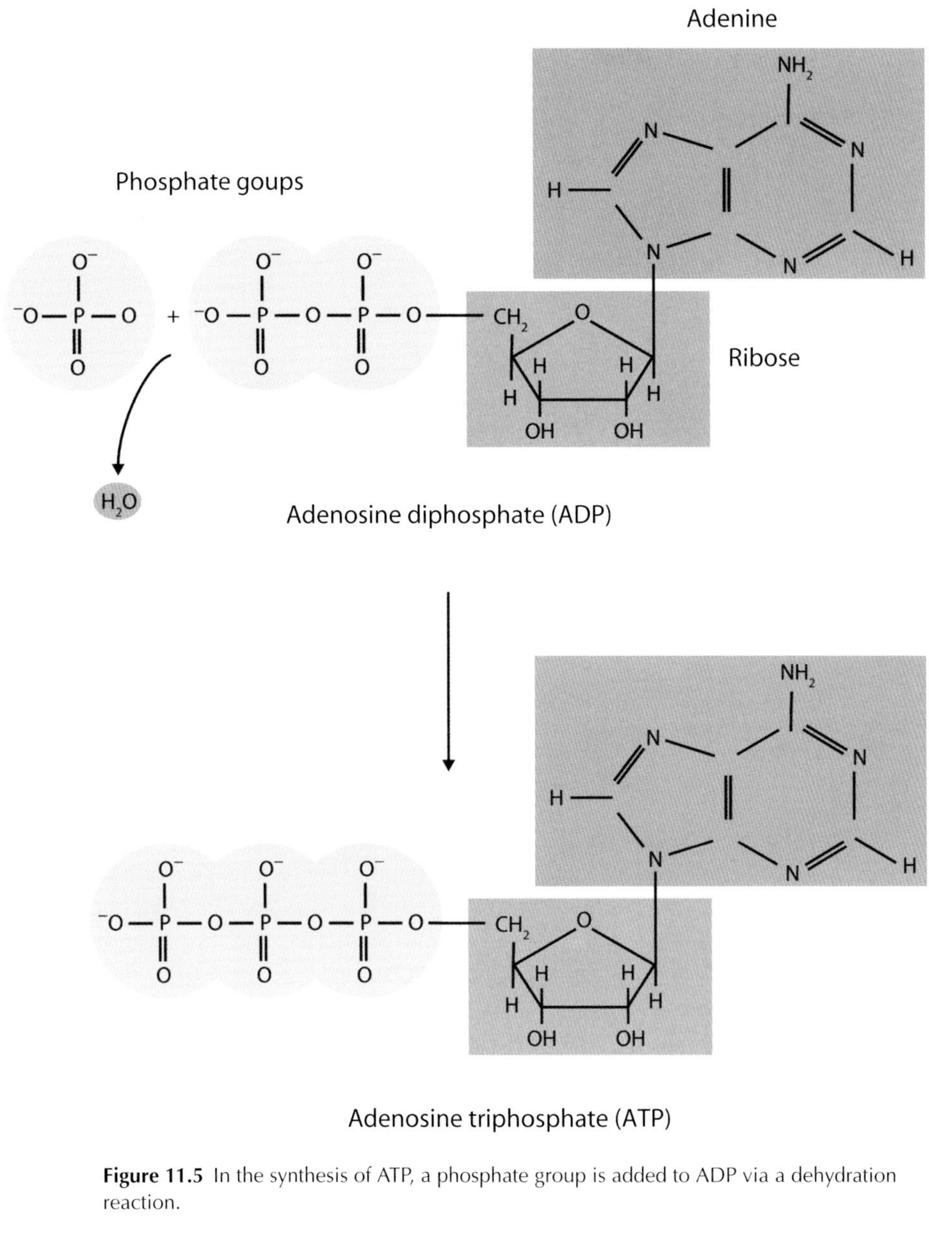

Figure 11.5 In the synthesis of ATP, a phosphate group is added to ADP via a dehydration reaction.

glycolysis, some of the energy in the triose phosphates is used to reduce NAD^+ to NADH and add a second inorganic phosphate to glyceraldehyde-3-phosphate. One of these phosphates is used to synthesize ATP from ADP, and in subsequent steps, a second ATP is formed. Therefore, if the substrate of glycolysis is considered to be one molecule of glucose or fructose, the net products are two molecules of ATP, two molecules of NADH, and two molecules of pyruvate.

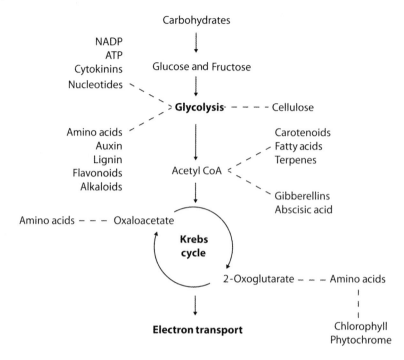

Figure 11.6 Respiration is integrated with the biochemistry of the cell and does not take place in isolation. The intermediates of respiration are used in the synthesis of many other cellular metabolites.

The main task that is accomplished by **glycolysis** is the breakdown of glucose, a 6-carbon compound, to two molecules of the 3-carbon compound pyruvate, the molecule that feeds into the Krebs cycle. The NADH produced in glycolysis must be actively transported across the mitochondrial inner membrane to be utilized to form ATP, and this active transport of NADH itself requires energy. The major release of energy from cellular respiration occurs in the steps that follow glycolysis.

Before the process of photosynthesis evolved, considerably less oxygen was available for aerobic (oxygen-based) respiration, the main source of energy (ATP) for metabolism. Therefore, glycolysis was the primary source of energy for metabolism. Similarly, under conditions of low oxygen availability, such as in waterlogged roots, the **electron transport chain** cannot go to completion, so the aerobic steps of respiration shut down and the pyruvate produced by the Krebs cycle is converted to ethanol or lactate (Figure 11.8). Glycolysis produces two molecules each of ATP and NADH for each molecule of glucose entering the pathway, but the NADH can be recycled to NAD^+ by the production of ethanol or lactate, which allows glycolysis to continue by freeing up NAD^+. The lactate and ethanol produced still retain most of the energy invested in the formation of carbohydrates during photosynthesis. Yeast routinely exploits this pathway, called **fermentation**,

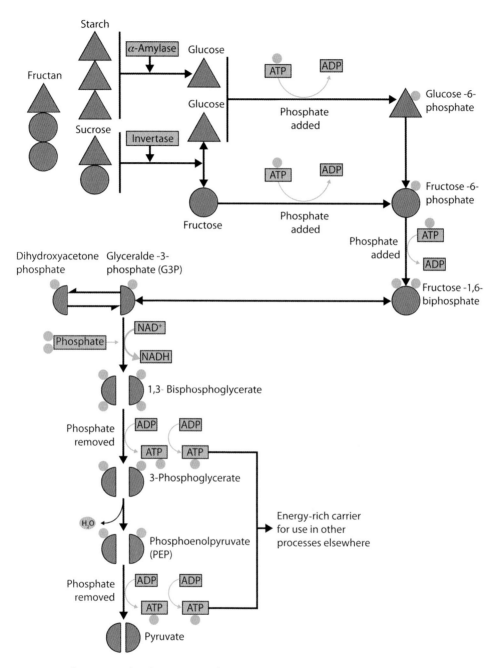

Figure 11.7 Glycolysis occurs in the cytosol. One molecule of glucose or fructose (6-carbon) is energized by the addition of phosphate bonds, and the products are two molecules of NADH, two molecules of ATP, and two molecules of pyruvate (2 × 3-carbon). Pyruvate is imported by mitochondria. (Diagram redrawn from Wilkins 1988, used with permission.)

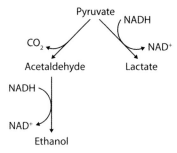

Figure 11.8 Under conditions of low oxygen, such as waterlogging of roots, the Krebs cycle and electron transport chain do not occur. If NAD^+ is not available, even glycolysis cannot continue. In low-oxygen conditions, fermentation uses the NADH produced in glycolysis to oxidize NADH and produce either ethanol (alcoholic fermentation) or lactate (lactic acid fermentation) freeing NAD^+ for glycolysis.

in baking and brewing. Products that contain ethanol, such as wine, will spoil after opening because bacteria will extract some of the energy left by fermentation by converting the ethanol to acetic acid (i.e., vinegar). The energy yield from fermentation of glucose (2 ATPs) is much less than the yield from aerobic respiration (36 ATPs). Fermentation, like glycolysis, takes place in the cytosol.

Mitochondria, the site of aerobic respiration, are organelles with two membranes—an outer membrane that is relatively nonselective and an inner membrane that is highly specialized. The Krebs cycle takes place in the inner matrix of the mitochondrion, the aqueous solution that occupies the space inside the inner membrane (Figure 1.13).

When pyruvate is transported from the cytosol into the inner matrix of the mitochondrion (Figure 11.9), a molecule of CO_2 is removed and a molecule of NADH is generated. A cofactor, coenzyme A (CoA) is added, converting the remaining 2-carbon molecule to acetyl-CoA. The bond between the 2-carbon acetyl group and CoA is high in energy, and is used to carry out the first step of the Krebs cycle.

The Krebs cycle

The **Krebs cycle** takes place in the inner matrix of mitochondria. Reducing power (**NADH** and **FADH$_2$**) is created in the Krebs cycle and subsequently used in the last step of respiration, electron transport, to make ATP. In the first step of the Krebs cycle, the 2-carbon acetyl group from acetyl CoA is added to OAA (oxaloacetic acid), a 4-carbon compound, to make the 6-carbon compound citric acid (Figure 11.9). The cofactor (CoA) is recycled and used for the conversion of the next pyruvate. Citric acid is composed of three carboxylic acids, and the two alternative names for the Krebs cycle are the TCA (tricarboxylic acid) cycle and the citric acid cycle (Figure 11.10).

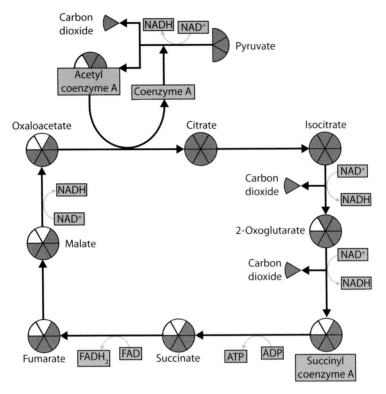

Figure 11.9 Pyruvate feeds into the Krebs cycle. The three carbons of pyruvate are released as CO_2, and for every two pyruvate molecules, eight NADH and two $FADH_2$ are produced, and two molecules of ATP are produced. (Diagram redrawn from Wilkins 1988, used with permission.)

Figure 11.10 The Krebs cycle is also referred to as the citric acid cycle and the tricarboxylic acid cycle. Citric acid, the molecule to which the 2-carbon acetyl group is added in the first step of the Krebs cycle, has three carboxylic acids.

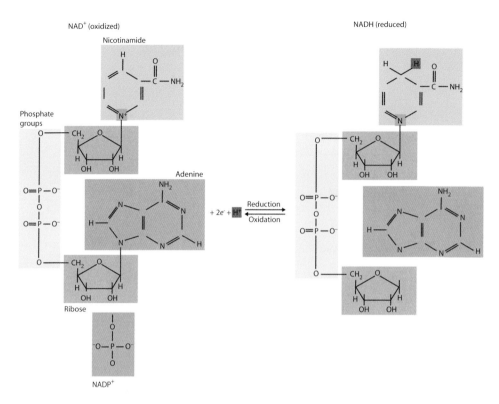

Figure 11.11 In the formation of NADH, two electrons and a proton are added to NAD$^+$. The reducing power of this molecule and FADH$_2$ can be used to sequester protons as they move through the electron transport chain. A similar molecule, NADHP, is formed from NADP$^+$ (orange) in photosynthesis.

As a molecule of citric acid is processed by successive steps of the Krebs cycle, two more carbons are lost as CO$_2$, one at a time, each time generating a molecule of NADH (Figure 11.11). In the second of these reactions, CoA is added to the product and this energy is subsequently used to synthesize a molecule of ATP. A molecule of **FADH$_2$** (flavin adenine dinucleotide) is produced in the Krebs cycle and can be used in the electron transport chain of cellular respiration along with NADH. In the last step of the Krebs cycle, OAA is regenerated and a third molecule of NADH synthesized. Because glucose or fructose is the source of two pyruvates for the Krebs cycle, the net gain in energy and reducing power per glucose equivalent is the production of two ATPs, six NADHs, and two molecules of FADH$_2$. Therefore, the Krebs cycle is the step of respiration that is the primary source of reducing power in the form of NADH and FADH$_2$.

The electron transport chain

The high-energy electrons added to NADH and FADH$_2$ in the Krebs cycle (Figure 11.11) are used in a series of oxidation–reduction reactions

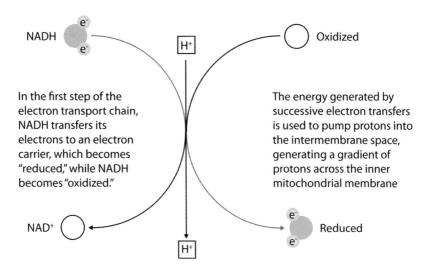

Figure 11.12 In an oxidation–reduction reaction, electrons (e^-) of one compound are transferred to another compound, along with charge-balancing protons (H^+). The compound losing the electrons is oxidized, and the compound receiving the electrons is reduced.

(Figure 11.12) where electrons from these compounds move through a series of protein complexes, termed the electron transport chain (Figure 11.13). The protein complexes of the electron transport chain are located either on or in the inner membrane of the mitochondrion (Figure 11.14) and most of these contain iron or iron and sulfur. The membranes of mitochondria, like the plasma membrane, are made up of lipids and proteins, and many of the proteins are involved in the electron transport chain. As in photosynthesis, the movement of electrons through the electron transport chain is coupled

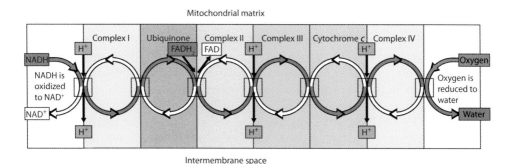

Figure 11.13 The electrons from NADH or $FADH_2$ reduce enzyme complexes in the electron transport chain. Each reduced molecule reduces the next enzyme complex, and in the process, protons are pumped into the intermembrane space. The final receptor for the electrons is oxygen, producing water. (Diagram redrawn from Wilkins 1988, used with permission.)

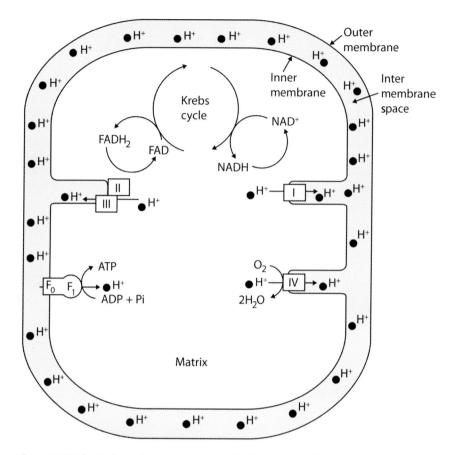

Figure 11.14 The Krebs cycle occurs in the matrix of the mitochondrion, inside the inner membrane. The reducing power of NADH and FADH$_2$ is used in the electron transport chain to pump protons (H$^+$) into the intermembrane space (yellow). The flow of protons back into the inner matrix is through ATP synthase. This flow causes a subunit of ATP synthase to rotate, providing the energy to add a phosphate to ADP.

to the pumping of protons (H$^+$) across a membrane; in this case from the inner matrix of the mitochondrion into the intermembrane space between the inner and outer membranes. Sequestering of H$^+$ is similar to that which occurs in photosynthesis, with the electrons moving to the outside of the membrane and eventually returning to the inner matrix.

ATP synthesis in respiration is driven, as it is in photosynthesis, by the movement of protons back across a membrane (Figure 11.15). In this case, protons move back into the inner matrix of the mitochondrion. The term **oxidative phosphorylation** refers to the series of oxidation–reduction reactions of electron transport that utilize the reducing power of NADH and FADH$_2$ and result in the phosphorylation (the addition of a phosphate group) of ADP, producing ATP.

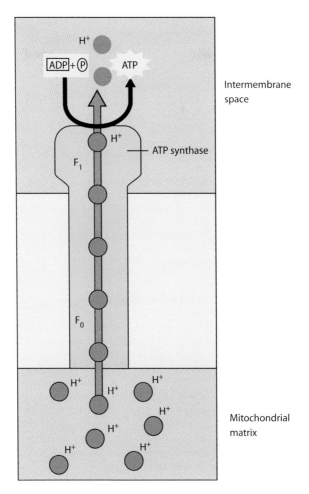

Figure 11.15 Protons sequestered in the intermembrane space flow back into the matrix through ATP synthase, adding a phosphate group to ADP. (Diagram redrawn from Wilkins 1988, used with permission.)

Molecular Motors Synthesize ATP

ATP is synthesized in the photochemical steps of photosynthesis and used in the biochemical steps of photosynthesis, and ATP is also synthesized—using the energy invested in organic molecules—in the last phase of respiration. In both cases, protons (H^+) are sequestered across a membrane, driven across by the energy released by oxidation–reduction reactions in an electron transport chain. The protons flow back through an integral (membrane-spanning) protein complex, driving the synthesis of ATP from the substrates ADP and inorganic phosphate.

The structure of ATP synthase is complex and the enzymatic mechanism that was proposed for its enzymology was novel—"rotational catalysis," in which the shape of active sites on the enzyme complex is altered by the rotation of a central protein subunit (Boyer 1993). Many lines of evidence supported both the structure and the

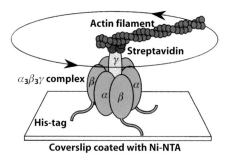

Figure 11.16 The experiment used to demonstrate that rotation within the enzyme is the mechanism for the synthesis of ATP: the F_1–γ subcomplex of ATP synthase was immobilized upside down on a glass plate, and a fluorescent actin filament many times larger than the protein complex was attached to the γ subunit, which functions like a rotor shaft and which would normally protrude into the F_0 component in the membrane. The actin filament was labeled with fluorescent dyes that allowed movement to be visualized when viewed under UV light. (Figure from Noji et al. 1997, used with permission.)

proposed mechanism, but an elegant experiment by scientists from the Tokyo Institute of Technology and Keio University in Yokohama, Japan, provided confirmation of rotation that was visible at the light microscope level (Noji et al. 1997).

The ATP synthase enzyme is composed of two "motors," F_0 and F_1 (Figure 11.15), that share a rotor shaft (the γ subunit). The F_0 subcomplex is embedded in the membrane and the F_1 subcomplex is composed of three α and three β subunits (Figure 11.16). The flow of protons through F_0 causes the γ subunit to rotate within the F_1 subcomplex, successively changing the conformation of the β subunits of F_1 and resulting in the synthesis of one molecule of ATP for each 120° rotation (Noji and Yoshida 2001). The γ subunit of the ATP synthase enzyme is able to rotate in the opposite direction, too. When the shaft rotates clockwise (when viewed from the membrane), ATP is synthesized, but when it turns counterclockwise, the same protein complex can hydrolyze ATP and pump protons across the membrane in the opposite direction, the mechanism used in the active uptake of nutrient ions (Figure 8.4).

To visualize the activity of ATP synthase, researchers attached the F_1–γ subcomplex of ATP synthase to a glass cover slip (Figure 11.16; Noji et al. 1997). To do this, the subcomplex was expressed in the bacterium Escherichia coli to have 10 His (the amino acid histidine) residues attached to each of the three β subunits. The His tags were able to stick to a Ni-NTA (nickel ion-nitrilotriacetic acid) coated glass cover slip. In further steps, a cofactor, biotin was attached to both the γ subunit and a fluorescently labeled actin filament. Biotin is strongly attracted to the protein streptavidin, so the actin filament and the γ subunit could be linked through streptavidin

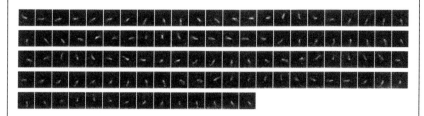

Figure 11.17 Successive photographs showing the counterclockwise movement of the actin filament attached to the γ subunit of an F_1–γ subcomplex of ATP synthase, indicating ATP hydrolysis. (Figure from Noji et al. 1997, used with permission.)

(Figure 11.16). The actin filament was more than 100 times the length of the protein complex, so much larger than the ATP synthase F_1 protein complex that it has been likened to a person waving a flag that is several hundred meters long (Taiz and Zeiger 2006). The F_1–γ subcomplex plus actin was viewed using UV light, which caused the fluorescent dyes on the actin filament to fluoresce. When ATP was added to the F_1–γ subcomplex of ATP synthase attached to the glass cover slip, the γ subunit with attached actin filament was photographed turning counterclockwise, the direction indicating ATP hydrolysis (Figure 11.17).

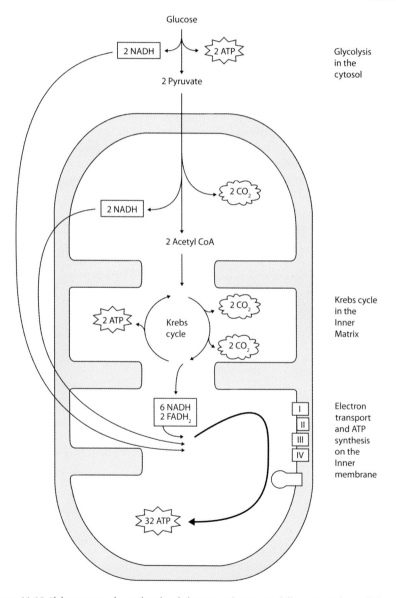

Figure 11.18 If the energy of a molecule of glucose or fructose is fully extracted via all the three steps of respiration, 32 molecules of ATP are produced from NADH and $FADH_2$, plus 2 ATP from glycolysis and 2 from the Krebs cycle.

The molecule that accepts the electrons as the last step of the electron transport chain is oxygen (O_2) (Figure 11.13). Each atom of oxygen is reduced to two molecules of water (H_2O) by the addition of four electrons from the electron transport chain (where each carry a negative charge), along with four positively charged protons (H^+) from the mitochondrial matrix (Figure 11.14). It is at this step that respiration uses oxygen, turning it into water. In our own bodies, while we exchange CO_2 from respiration for O_2 to support respiration in our lungs, these compounds are produced and used, respectively, at the cellular level in the Krebs cycle and the electron transport chain.

In summary (Figure 11.18), the steps of respiration include glycolysis, which occurs in the cytosol; the Krebs cycle, which occurs in the inner matrix of the mitochondria; and the electron transport chain, which occurs on the inner membrane of mitochondria. If a molecule of glucose is followed through the entire biochemical pathway, 2 more ATPs are synthesized than are used in glycolysis, 2 are synthesized in the Krebs cycle, and as many as 28 in the electron transport chain, for a total of 32 ATPs.

Each of the steps of respiration is inefficient, so much of the energy originally invested in glucose is not captured in respiration. Only a small percentage of the solar energy falling on a leaf is captured as carbohydrates in photosynthesis, and eventually *all* energy captured from the sun is expended as heat. Therefore, the constant replenishment of the food chain from solar energy into photosynthesis is fundamental for our survival.

Chapter 12
Environmental regulation of plant development

Plants continuously respond to stimuli in their environment such as temperature, light, gravity, and mechanical disturbance; many of these responses are controlled through plant hormones. However, seasonal changes in the environment of temperate plants impose the most difficult challenges of surviving winter and reproducing effectively, and therefore have profound effects on growth, dormancy, and reproduction.

The most important response for the long-term survival of a plant species is flowering, seed production, and the establishment of seedlings. Some plants flower early in the growing season, some later in the year, and others flower all season long. Our understanding of this set of responses of plants began with experiments on flowering done by Wightman Garner and Henry Allard in the early years of the twentieth century at the USDA Bureau of Plant Industry (now the Agricultural Research Service) in Beltsville, Maryland.

Photoperiodism

The general relationship between the light required for photosynthesis and plant growth was appreciated, but the climatic stimulus for flowering had not been determined in the early 1900s. There was an assumption that changing temperatures over the course of the growing season were likely to control flowering. Garner and Allard tested the effect of different levels of temperature, moisture, nutrition, and light intensity on the flowering of soybeans, but it was only when they varied the duration of the daylight period (**daylength**) that they found the environmental cue that influenced flowering (Garner and Allard 1920). They determined that when the soybean (*Glycine max*) cultivar "Biloxi" was seeded at intervals over a 5-month period from April through August, all the plants flowered within a *5-week* period from early September through early October (Figure 12.1). This result suggested that temperature and plant maturity were not the factors determining flowering

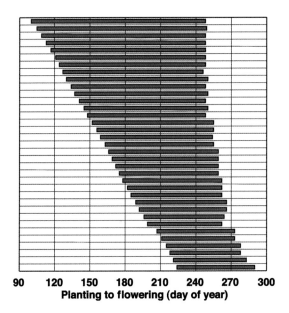

Figure 12.1 Seed of "Biloxi" soybeans were planted near Washington, DC approximately every 3 days from April through October. Each bar represents the length of time to flowering. (Data from Garner and Allard 1920.)

date. They also reported that a tall, large-leaf mutant of the "Maryland" cultivar of tobacco (*Nicotiana tabacum*) called "Maryland Mammoth" did not flower in the field in August but did flower consistently in late winter if transplanted to the greenhouse.

In further experiments where Garner and Allard put Maryland Mammoth plants in the dark for several hours, artificially shortening the length of the day, the plants produced flowers in the field in summer (Figure 12.2). They subsequently found that while some plants, like tobacco and soybean, flowered in response to daylengths that were shorter than some maximum, other plants, including many of the grasses and cereal grains, flowered in response to daylengths that were longer than some minimum. The terms "short-day" for plants flowering late in the season and "long-day" for plants flowering earlier were used to classify these responses. Plants that did not respond to daylength signals were classified as "**day-neutral**" plants. Garner and Allard termed the response of flowering to daylength "**photoperiodism.**"

The advantage for plants in using daylength to determine when flowering and subsequent seed production will occur is that change in daylength is the most reliable environmental signal available to foretell seasonal change. The Earth rotates on an axis through its north and south poles. If a line drawn from the Earth to the sun were considered horizontal, this axis is inclined 23.5° from vertical (Figure 12.3). As Earth revolves around the sun, the northern hemisphere is inclined away from the sun for half the year and toward the sun the other half. Daylength changes as the Earth revolves around the sun over the course of the year, and the maximum daylength occurs at the

Figure 12.2 (a) The "Maryland Mammoth" plants in the top panel were grown under 12-hour days and were beginning to form buds when this photograph was taken on August 19. (b) The "Maryland Mammoth" plants in the lower panel were grown under 7-hour daylengths over the same time period and seed pods had formed by August 19. (Figures from Garner and Allard 1920.)

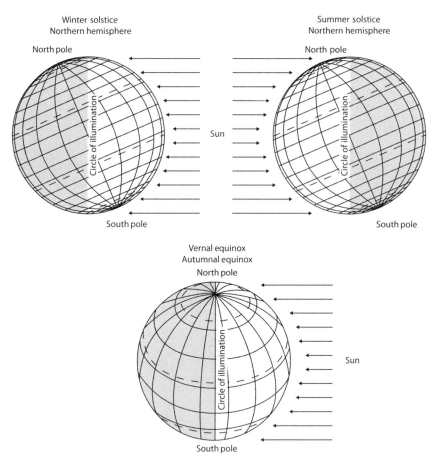

Figure 12.3 At the winter solstice (upper left), the northern hemisphere is at its greatest angle from the sun, making this the shortest day of the year. At the summer solstice (upper right), the northern hemisphere is angled most directly toward the sun, making this the longest day of the year. The equinox (meaning equal night; lower panel), which falls in mid-March and mid-September, are the days when the Earth's axis is vertical with respect to the sun, and this results in 12-hour days at all locations on Earth. (Figure from Trewartha and Horn 1980, reproduced with permission from The McGraw-Hill Companies.)

summer solstice. As latitude (distance from the equator) increases, daylength at the summer solstice increases (Figure 12.4). Plants adapted to higher and thus colder latitudes have shorter windows in summer to mature seeds; therefore, temperate plants tend to be long-day plants that flower in late spring.

As days lengthen in spring (Figure 12.5), long-day plants will flower when daylengths exceed a minimum. Short-day plants flower later in the summer, following the summer solstice. Scientists using cocklebur, a short-day plant, demonstrated that interrupting the day period with darkness had no effect on flowering, while interruption of the night period inhibited flowering (Figure 12.5). For cocklebur, it took only a 1-minute pulse of light during the night to undo the effect of the night period. What was eventually demonstrated

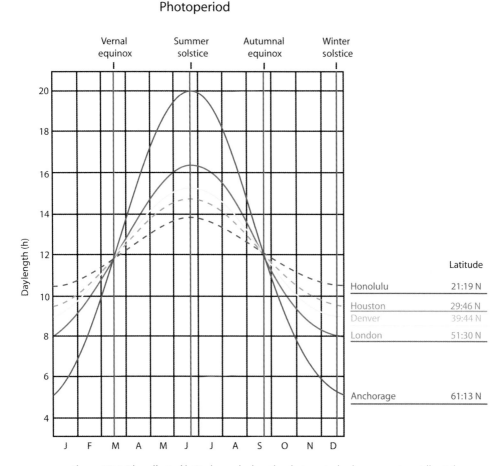

Figure 12.4 The effect of latitude on daylength: photoperiods change most rapidly at the equinox in spring and autumn; the greater the latitude, the greater the daylength at the summer solstice.

was that it is the length of the night rather than the length of the day that is the critical stimulus for flowering.

Phytochrome

By testing the effectiveness of specific wavelengths of light within or just beyond the visible spectrum on the germination of lettuce seeds, it was demonstrated that red light (wavelengths from 620 to 700 nm) particularly near 660 nm, promoted germination, while far-red light (wavelengths from 700 to 775 nm), with peak activity at 730 nm, had just the opposite effect (Figure 12.6). In addition to demonstrating the action spectrum for phytochrome, the same paper (Borthwick *et al.* 1952a, b) reported the remarkable discovery that germination was switched on if plants were exposed to red light (R),

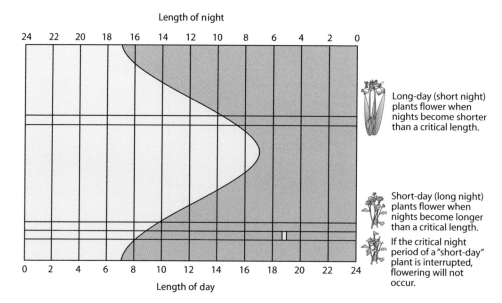

Figure 12.5 The effect of change in daylength (yellow) from the winter solstice (top) through summer and autumn on flowering of long-day and short-day plants. Short-day plants whose night period is interrupted with a short burst of light will not flower.

but switched off if plants were subsequently exposed to far-red (FR) light. The pattern remained the same when the sequence of R–FR–R–FR (on–off–on–off) was repeated: the last exposure of seeds to red or far-red light determined germination (R) or suppression of germination (FR). It is also notable that the authors made a novel speculation in this same paper: that a single pigment changing from one form to another could explain the observed results.

It was later confirmed that the pigment absorbed red light, causing it to be converted to a form that absorbed light in the far-red part of the spectrum. When the far-red form of the pigment absorbed far-red light, the pigment was converted back to the red-absorbing form. The same group that determined the action spectrum for phytochrome was able to isolate the pigment 7 years later (Butler et al. 1959). By demonstrating that changes in this pigment, named **phytochrome**, corresponded to red and far-red light responses in plants, phytochrome was shown to be the agent for these responses. The two isomers (alike in chemical formula) of phytochrome differ only in the orientation of the last ring of the pigment molecule.

The active form of the pigment is phytochrome far-red (P_{FR}). Phytochrome is synthesized as phytochrome red (P_R), and absorption of red light or direct sunlight, which is enriched in red light, converts P_R to P_{FR} (Figure 12.7). At night, P_{FR} slowly reverts to P_R, which is the inactive form. As days become longer in the spring, the **critical photoperiod** for a long-day plant is reached when nights become too short for all the P_{FR} produced during the day to revert to P_R.

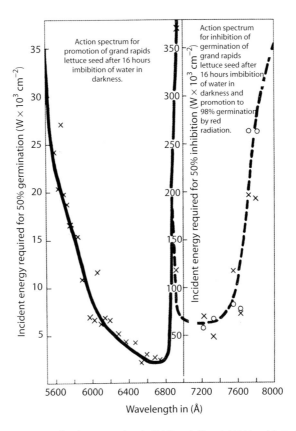

Figure 12.6 Action spectra for the promotion (solid line, left) or inhibition (dotted line, right) of lettuce seed germination. On the x-axis, 10 Å is equal to 1 nm. The results are reported as light energy required to evoke a response, so the lower the value, the more effective the light: light near 6,600 Å (660 nm) promotes germination, while inhibition is caused by light near 7,300 Å (730 nm). (Figure from Butler 1964, used with permission.)

For a long-day plant, a long day and a short night result in synthesis of enough P_{FR} to cause flowering. However, if a long-day plant is in an artificial environment with both long days *and* long nights, it will not flower because much of the active form of P_{FR} is converted back to P_R during a long night.

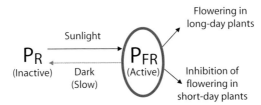

Figure 12.7 Phytochrome red (P_R) absorbs red light (620–700 nm), which converts it to phytochrome far-red (P_{FR}), the active form. At night, P_{FR} gradually reverts to P_R; if P_{FR} absorbs far-red light (700–775 nm), it will rapidly be converted to P_R. In long-day plants, accumulation of P_{FR} results in flowering; in short-day plants, accumulation of P_{FR} causes inhibition of flowering.

In short-day plants, P_{FR} has the opposite effect—it *inhibits* flowering, so as days become shorter in late summer and more P_{FR} is converted back to the inactive P_R form during the longer night period, a point is finally reached after which the inhibition of flowering by P_{FR} is removed, and flowering occurs. When nights are short earlier in the season, these plants are not physiologically mature enough to flower.

The interconvertibility of phytochrome from the red to the far-red form allows plants to time their flowering so that other less-reliable environmental factors, such as temperature, are likely to be optimal for completion of their lifecycle. It also allows seeds to germinate when there is sufficient time for them to become established before winter.

Florigen

How is the response of phytochrome in plants translated into flowering? For many years, there was evidence for the existence of a substance, formed in the leaves and translocated through the phloem to the apical or axillary meristems, which transformed these organs from vegetative to reproductive (flowers and then fruit). If the apical meristem of a short-day plant is the only organ exposed to short days, flowering does not occur (Figure 12.8). If the leaves, or even just one leaf, of the plant are exposed to short days, flowering will occur. Even when the leaf exposed to short days is on a separate plant, connected only by a graft, the substance causing flowering will be translocated (Figure 12.9). The evidence was so strong for the existence of this translocated substance that it was named "**florigen**" before it was isolated.

In 2007, florigen was identified in *Arabidopsis*, which is a long-day plant, to be FT protein (Corbesier *et al.* 2007), and in rice, a short-day plant, to be Hd3, a protein (Tamaki *et al.* 2007) which differs only slightly from FT protein. The relationship between phytochrome and these proteins is illustrated in Figure 12.10. Long days (upper panel) increase the leaf content of phytochrome far-red (P_{FR}). Phytochrome red is located in the cytosol, but phytochrome far-red moves into the nucleus where it initiates gene expression. In long-day plants, accumulation of P_{FR} results in the synthesis of the floral stimulus mRNA and its protein, and in short-day plants, this results in the synthesis of an inhibitor of the floral stimulus. Short days (lower panel) result in an insufficient level of P_{FR} to stimulate flowering in long-day plants, and in short-day plants, without the inhibitor of flowering, the floral stimulus is synthesized. Florigen, or FT protein, is synthesized in the phloem of leaves and translocated through the phloem to the apical meristem, where it stimulates a biochemical pathway that transforms the vegetative meristem to a reproductive meristem (Figure 12.11).

Other phytochrome-mediated responses

Seeds of some plants, many of them weeds, use phytochrome to acquire information about the amount of shading and therefore competition to

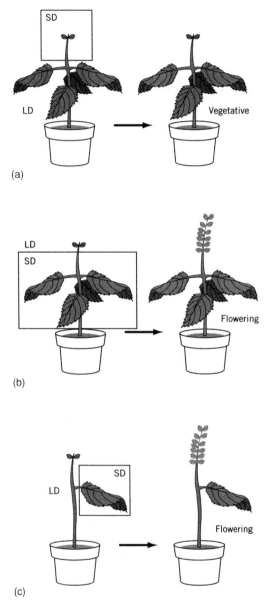

Figure 12.8 Demonstration of the site of the response to photoperiod. (a) When only the apical meristem of a short-day plant (*Perilla*) was exposed to short days, the plant remained vegetative. (b) When the leaves were exposed to short days and the apical meristem was not, the plant flowered. (c) When a single leaf was exposed to short days, flowering occurred. SD, short days; LD, long days. (Drawing from Hopkins 1995, used with permission.)

determine whether they should germinate or not. Sunlight contains slightly (10%) more red than far-red light (Figure 12.12), but the photosynthetic pigments in leaves of plants absorb or reflect much of the light in the visible portion of the spectrum (400–700 nm; red is 620–700 nm), while light in the far-red range (700–775 nm) is transmitted through leaves. Therefore, a

Figure 12.9 Demonstration of the translocation of the floral stimulus from one plant to another. When plants were grafted together and only one plant was exposed to short days, all plants were induced to flowering. SD, short days. (Drawing from Hopkins 1995, based on Chailakhian 1961, used with permission from both sources.)

seed lying near the surface of the soil underneath a canopy formed by the leaves of other plants will receive light greatly enhanced in P_{FR} light (Figure 12.13) compared to a seed germinated in full sunlight.

Any P_{FR} in a seed under a canopy would absorb far-red light which would convert the P_{FR} to P_R. A seed under a leafy canopy would therefore contain more phytochrome in the P_R (inactive) form than in the P_{FR} (active) form, and would be less likely to germinate. The effect of red and far-red light on the germination of "Grand Rapids" lettuce seed (Figure 12.14) illustrates the power of this mechanism.

Stem elongation has also been shown, in some plants, to respond to the ratio of red to far-red light that plants receive. Because red light wavelengths are absorbed from the light that passes through leaves, the red to far-red ratio (R:FR) reaching an elongating stem within a canopy would be decreased. Absorption of far-red light converts P_{FR} back to P_R and phytochrome in the far-red state inhibits stem elongation. Therefore, the lower the ratio of R:FR in the light the plant receives, the lower P_{FR} will be in the stem and the longer the stem will become in order to carry leaves to the top of the canopy where light for photosynthesis is likely to be more abundant (Figure 12.15).

When pea seeds are germinated in the dark, the epicotyl will continue to elongate and remain light yellow in color (Figure 12.16). The leaves will not expand and the epicotyl will retain its hook. This **etiolation** following seed germination in the dark functions to allow deeply planted seeds to reach the soil surface. When exposed to light, the epicotyl or hypocotyl hook will uncurl, or the coleoptile of grass seedlings will stop elongation, and leaves will expand and turn green. The action spectrum for stem elongation shows that it is a phytochrome response, but also shows a response in the UV and blue regions of the light spectrum. There are indeed other light

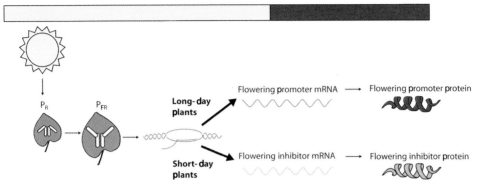

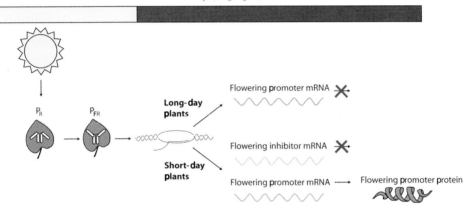

Figure 12.10 Under long days (upper panel), phytochrome far-red (P_{FR}) accumulates. P_{FR} stimulates gene expression and protein synthesis. The new protein produced in long-day plants causes flowering, and the new protein produced in short-day plants causes flowering to be inhibited. Under short days (lower panel), P_{FR} reverts to P_R at a sufficient rate to keep P_{FR} levels low. This is insufficient to promote flowering in long-day plants or to inhibit the production of the floral stimulus in short-day plants. Without inhibition, a new protein that promotes flowering forms.

receptors involved in plant growth, development, and function in addition to phytochrome.

Plant environmental responses requiring cold temperature

Many perennial plants that are considered long-day plants also require **vernalization**, or exposure to chilling that leads to the ability to flower (Chouard 1961). Vernalization at the molecular level is thought to act by switching off

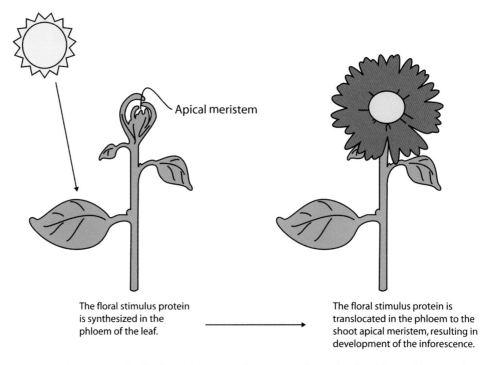

Figure 12.11 The floral stimulus protein florigen is synthesized in the phloem of leaves and translocated to the floral meristem, which is transformed from vegetative to reproductive.

a gene that inhibits flowering. Vernalization accelerates flowering in winter annual varieties of small grains such as wheat, oats, barley, and rye that are planted in the fall and produce seedheads the following spring. Decreasing temperatures and shortening days in autumn bring about biochemical

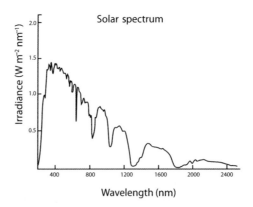

Figure 12.12 The solar spectrum of electromagnetic energy contains about 10% more red (620–700 nm) than far-red (700–775 nm) light. Therefore, in plants exposed to direct sunlight, phytochrome red (P_R) will be converted to phytochrome far-red (P_{FR}).

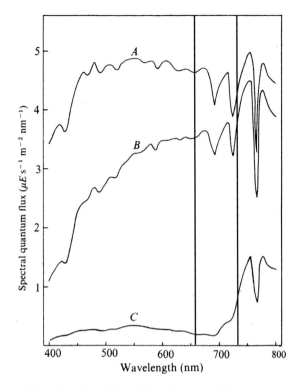

Figure 12.13 Daylight within the visible and far-red portions of the spectrum (400–800 nm) measured in late morning in early summer: (A) light above a winter wheat canopy, (B) at ground level in a sun fleck, and (C) at ground level in the shade of the wheat leaf canopy. Lines have been added to the figure near the most effective red and far-red wavelengths. (Figure from Holmes and Smith 1975, used with permission.)

changes in the meristem that are needed for the formation of flowers (termed **induction**), but physical changes are not apparent at the apical meristem (Figure 12.17). For long-day plants, including most temperate perennial grasses, increasing temperatures and lengthening days the following spring do result in physical changes—the **initiation** of a reproductive meristem that quickly develops into a seedhead.

In biennials, vernalization is required for flowering; biennials such as carrots, celery, turnips, and beets will only flower in their second year. Cold treatment of 1–3 months is required to vernalize most plants, but once a plant or even a seed has been vernalized, it retains the vernalized state but cannot pass it on to the next generation.

Some seeds require a period of cold temperature, termed **stratification**, before they will germinate (Figure 12.18). This can be supplied artificially by allowing seeds to absorb water, then storing the imbibed seeds in the cold (1–3°C) for 1–3 months, depending on the species. Similarly, some bulbs and reproductive structures of perennial plants will only produce flowers after being stored in cool temperatures for a given length of time. These treatments simulate the natural conditions of a winter period.

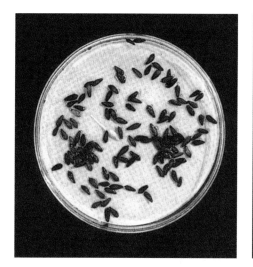
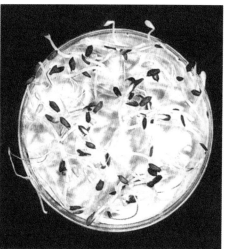
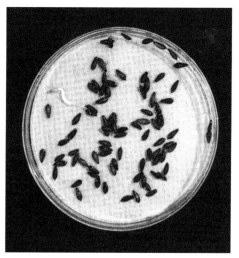

Figure 12.14 The lettuce (*Lactuca sativa*) variety "Grand Rapids" is light-sensitive and is used to illustrate the effects of phytochrome on seed germination. Seeds were placed on wet filter paper at 30°C. Those on the top left were kept in darkness, those on the top right were exposed to red light for 5 minutes followed by darkness, which caused them to germinate. The seeds in the lower panel were exposed to red light for 5 minutes, to far-red light for 5 minutes, then darkness. The exposure to far-red light following exposure to red light prevented germination. (Photograph from Wilkins 1988, used with permission.)

Winterhardiness of perennial plants

Dormancy is a temporary cessation of growth that can occur in response to changes in the climate such as excessive temperatures or drought. In some plants, exposure to cold temperatures is required to break dormancy, such

Figure 12.15 The effect of 14 days of exposure to four ratios of red to far-red light on stem elongation in lamb's quarters (*Chenopodium album*). The higher R to FR ratio on the left is about twice the R:FR of sunlight; in shade, the R:FR can drop to less than 0.1 (Smith and Morgan 1981, used with permission; photograph courtesy of Harry Smith)

as the dormancy of a flower bud (Figure 12.19). In contrast to vernalization, which is necessary for the *development* of the floral meristem, the chilling needed to break flower bud dormancy is acting on an already-developed organ or its primordium in which growth has been arrested. Fall dormancy is a response to shortening daylengths and decreasing night temperatures—changes in the environment that predict rather than measure harsh environmental conditions. In dormant plants, cells needed for resumption of growth remain alive and function throughout the dormant period.

A critical response of plants for winter hardening or even resistance to chilling stress changes in membranes, including the plasma membrane, the tonoplast, and the membranes of mitochondria and chloroplasts that allow them to continue to function as temperatures decrease. Membrane function requires fluidity of the lipid component of membranes, which also affects the functioning of proteins and protein complexes such as the ion channels that are located within membranes.

Fats that are highly saturated, such as lard, are solid at room temperature, while fats that are liquid at room temperature, such as olive oil, are not fully saturated. Membrane lipids that function well during the heat of summer

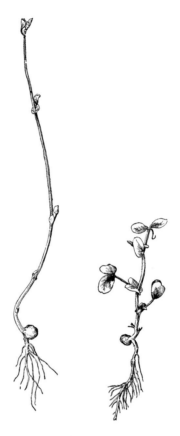

Figure 12.16 Diagram of germinating pea (*Pisum sativum*) seeds that developed in the dark (left) or in the light. (Drawing from Dugger 1924.)

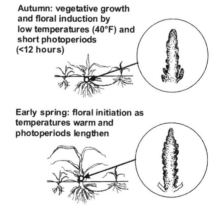

Figure 12.17 Vernalization is the competence to flower in the spring that is acquired by exposure to the cold temperatures of winter. Plants of perennial or winter annual grasses will be in the vegetative stage in autumn when they are exposed to decreasing daylengths and cold temperatures (induction). In spring, exposure to increasing daylengths and temperatures result in the initiation of a reproductive meristem. (Drawing from Volenec and Nelson 2003, used with permission.)

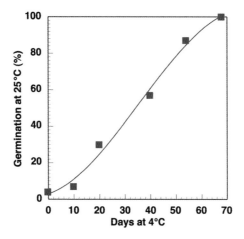

Figure 12.18 Effect of the length of stratification (days under moist conditions at 4°C) on germination of apple (*Malus*) variety "Crawley Beauty" seeds 5 days after transfer to 25°C. (Data from Luckwill 1952.)

contain saturated fatty acids that would crystallize at low winter temperatures, so these saturated fatty acids must be desaturated to increase the level of unsaturated fatty acids in membranes, as cells prepare for winter to allow the lipid bilayer to remain fluid at low temperatures.

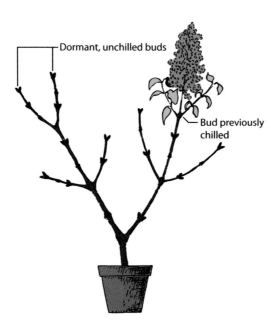

Figure 12.19 Dormancy of floral meristems can sometimes be broken by exposure to chilling, as illustrated here. Only the lilac bud exposed to cold developed into an inflorescence. (Figure 26.16, p. 537, from John W. Kimball 1983. Copyright © (1983) by Addison-Wesley Publishing Company, Inc. Reprinted by permission of Pearson Education, Inc.)

Figure 12.20 Leaves of a young shoot of alfalfa (*Medicago sativa*) after an autumn overnight frost. Ice that has formed in the apoplast (cell walls and intercellular spaces) can be reabsorbed by cells, as long as the cells have not been damaged by the formation of ice in the protoplast.

Plants can be injured in winter by the formation of ice crystals inside cells. Frost that occurs on the surface of leaves (Figure 12.20) is propagated through stomata into intercellular spaces. Ice formation inside cells is avoided by moving ions such as potassium (K^+) into the intercellular space, thereby removing water from cells as ice formation occurs in the cell wall and air spaces. As the water in the apoplast freezes, more liquid water is drawn from the protoplast into the apoplast (Figure 12.21), while proteins and sugars become more concentrated in the cytosol. As external temperatures increase following frost, cells reabsorb the ions and water from the apoplast.

Winter-hardy plants that are exposed to freezing are able to move water into the apoplast quickly enough to avoid ice crystal formation inside the protoplast. They can survive freezing, while the membranes of non-hardy plants

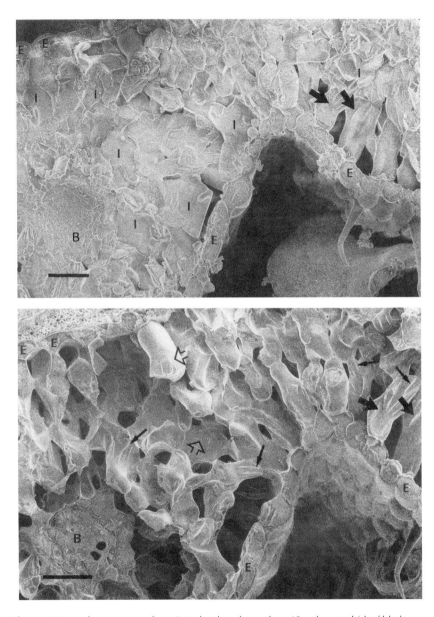

Figure 12.21 In the upper panel, a winter-hardened cereal rye (*Secale cereale*) leaf blade section was cooled at 2°C/h to −3.3°C/h, simulating the natural occurrence of frost. The intercellular spaces are filled with ice crystals (I), and cells are collapsed. In the lower panel, ice has been driven off as water vapor, allowing cells to be seen. The filled arrows in both panels point to the same cells. Some mesophyll cells are more collapsed than others (open arrows); cells remain attached to one another, causing ridges to form in walls (small filled arrows). B, vascular bundle; E, epidermis. (Photograph from Pearce 1988, used with permission.)

214 Structure and Function of Plants

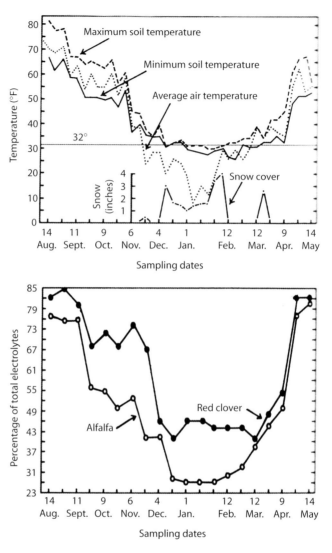

Figure 12.22 Winterhardiness increases slowly in fall and decreases rapidly in spring in the perennial forages alfalfa and red clover. Upper panel: change in air and soil temperature and snow cover during the winter of 1956–1957. Lower panel: leakage of electrolytes from intact roots and crowns of seedlings as a percent of total homogenate electrolytes. Higher electrolyte leakage indicates greater damage by freezing. By January, damage to tissue from freezing was minimal. Plants were dug from the field at 2-week intervals, frozen at −8°C for 4 hours, and suspended in distilled water for 20 hours at 2°C to determine electrical conductance of leachate from intact plants. The plants were then homogenized and electrical conductance of the filtered homogenate (total) determined. (Figure from Jung and Smith 1961, used with permission.)

will be damaged. This injury can be detected as salts and sugars (electrolytes) leaking from damaged cells (Figure 12.22). Winterhardiness is quickly lost in the spring, and therefore much "winter" damage often occurs from late freezes in spring after hardening is lost.

Chapter 13
Hormonal regulation of plant development

Plants are anchored in place by their roots and have developed tools to adapt and respond to their environment. In the previous chapter, some responses allowing plants to survive the seasonal fluctuations of temperate climates were examined, such as the ability to survive winter and to flower and germinate seeds under optimal conditions. In this chapter, the use of plant growth regulators to read and respond to the environment is discussed, and the topic of the last chapter will be plant defenses expressed as secondary metabolites.

Plant hormones are chemical messengers that are effective in very small quantities. Plant cells that respond to a particular hormone will have a receptor specific for that hormone which makes the tissue sensitive to the hormone. Hormones are usually transported from the site of synthesis to another location in the plant, where they elicit a physiological response that depends on their concentration and the sensitivity of the plant tissue. Often the effect produced results from the combined influence of two or three plant hormones.

There are five major groups of plant hormones classified by their general ability to promote or inhibit growth:

1. Auxins (most often IAA)—growth promoters
2. Gibberellins (GAs)—growth promoters
3. Cytokinins—growth promoters
4. Ethylene—growth inhibitor
5. Abscisic acid (ABA)—growth inhibitor

There are a number of other compounds that are similarly able to modify plant growth and development, but these five have well-understood effects and in many cases, commercial applications.

Auxins

Plants may contain more than one auxin, but the most active naturally occurring auxin is indole-3-acetic acid, or IAA. **Auxins** promote plant growth primarily by stimulating cell expansion.

Location of IAA synthesis

IAA is synthesized in actively growing shoot or reproductive plant organs such as shoot apical meristems, young leaves, and young fruits, flowers, and seeds.

Transport of IAA

While other plant hormones or their precursors are passively transported in the xylem or the phloem, IAA is actively transported (meaning faster than just by diffusion) through parenchyma tissue, especially parenchyma cells associated with vascular tissue. IAA generally moves down through the plant, toward the base in shoots and toward the root tips in roots. Movement like this in a specific direction is termed **"polar" movement**. Auxin is exported from one cell to another via efflux carriers that are localized only on the lower side of cells (Taiz and Zeiger 2006). Passive auxin transport also occurs in the phloem.

Effects of IAA

Cell elongation

IAA contributes to the growth of cells in the stems of dicots and the coleoptile of grasses by loosening the cell wall. In general, the driving force for growth is the uptake of water, which exerts pressure on the cell wall (Figure 8.5). However, in order for the cell to enlarge, the cell wall has to be able to expand. IAA increases the extensibility of the cell wall by increasing the activity of H^+-ATPases (Figure 8.4). H^+-ATPases cause protons (H^+) to be pumped into the cell wall, interfering with the hydrogen bonds linking cell wall structural carbohydrates to each other, and in this way loosening the wall.

Tropisms

Tropisms are growth in response to a stimulus. For plants to grow toward a source of light (**phototropism**), growth must be greater on one side of the shoot than on the other (Figure 13.1). However, this could occur through

Figure 13.1 Phototropism is growth of a plant shoot toward unidirectional light.

growth inhibition on the lighted side or growth promotion on the shaded side. As long ago as the 1870s, Charles Darwin and his son Francis observed that the tip of the coleoptile of grass seedlings sensed the light from a window in a darkened room. If the tip of the coleoptile was covered or removed, the coleoptile did not grow toward light. However, if the growing region was covered while the tip was still exposed to light, growth toward light *did* occur. They also noted that growth toward the light occurred below the tip of the oat coleoptile. The Darwins reasoned that since the light was sensed at the coleoptile tip, while growth toward the light took place below the tip, the substance altering growth was being transported from the tip of the coleoptile to the growth region.

If the tips of the coleoptiles were cut off and placed on agar (a gelatin-like substance), the compound would move into the agar. It was subsequently demonstrated that growth toward light was caused by promotion of elongation on the side of the coleoptile away from the light, while growth on the lighted side was inhibited. Light caused the compound in the coleoptile

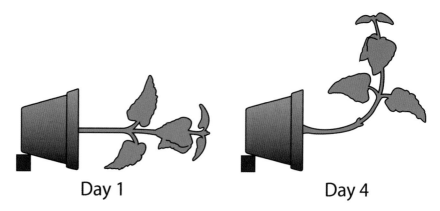

Figure 13.2 Gravitropism in the shoot is growth away from gravity.

tip that was responsible for growth to move away from the source of light. When the compound was isolated, it was identified as IAA. Light receptors in coleoptiles are sensitive to blue light, and changes in the receptors in response to blue light cause movement of auxin. The growth of a stem or coleoptile is promoted by the synthesis of auxin at the shoot apical meristem or coleoptile tip. The hormone is transported down, away from the shoot tip, and acts to promote elongation by stimulating H^+ efflux into the cell wall.

Gravitropism is a growth response to gravity. In shoots, growth is away from gravity (Figure 13.2) and is due to promotion of elongation by auxins. Auxin is redistributed in both the shoot tip and the elongation zone of the shoot in response to gravity. In the shoot, gravity is sensed by the shifting of amyloplasts to the lower side of cells if a shoot is turned from vertical to horizontal (Figure 13.3). However, root gravitropism (growth toward gravity) is caused by *inhibition* of growth by relatively high auxin levels in cells on the lower side of the root (Figure 13.4). The gravity-sensing amyloplast-containing cells of the root are in the root cap. Auxin from the shoot moves to the root tip through the parenchyma in the stele (vascular tissue). When it reaches the root cap, it circulates back up to the root elongation zone in the cortex and epidermis. In vertical roots, this auxin is equally distributed, but in a root reoriented to horizontal, the root cap directs most auxin to the lower cortex and epidermis. The excess concentration inhibits elongation on the lower side; root cell growth is inhibited by levels of auxin that would stimulate shoot growth.

Thigmotropism is the response of plants to touch (such as the growth of tendrils around a plant stake), and is also mediated by auxin.

Apical dominance

Pruning that removes an apical meristem stimulates the growth of branches from axillary buds. The inhibiting effect of the shoot apical meristem on the

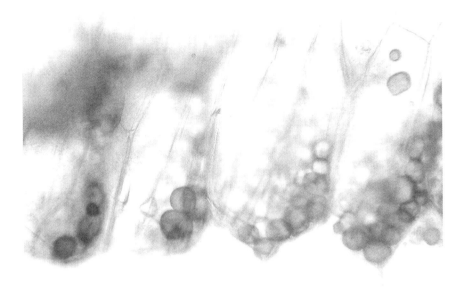

Figure 13.3 Cells sense gravity through the movement of amyloplasts, possibly by exerting pressure on the endoplasmic reticulum.

growth of axillary buds is termed **apical dominance**. In plants with strong apical dominance, branching near the shoot apical meristem is inhibited and plant formed is conical (Figure 13.5a). It is thought that the auxin produced by the apical meristem is only one of the factors interacting to inhibit the growth of axillary buds. When the apical meristem is removed and production of IAA is reduced, the auxin–cytokinin balance in the vascular tissue is altered. Abscisic acid is probably also involved. The branching pattern of plants that have weak apical dominance produces a spherical plant form (Figure 13.5b).

Formation of adventitious roots

When a leaf is excised from a plant and placed in a moist environment, the auxins produced by the leaf accumulate above the wound site. These relatively high levels of auxin, especially in the absence of the hormone cytokinin from the root, cause the plant to form roots at the cut. The importance of the normal polar movement of IAA downward through the xylem parenchyma becomes apparent in the generation of branch and adventitious roots (the normal permanent root system of grasses, shown in Figure 5.1a). The function of roots is to supply water and mineral nutrients from the soil to the shoot, and auxin triggers the differentiation of vascular tissue in these new roots. Auxins are often applied to shoot cuttings in the form of compounds such as RooTone® to encourage faster rooting.

220 *Structure and Function of Plants*

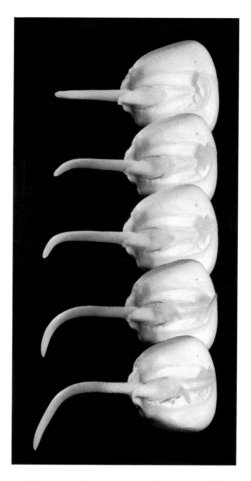

Figure 13.4 In roots, gravity is sensed by movement of amyloplasts in the root cap. Increased IAA on the lower side of the elongation zone inhibits growth, causing roots to reorient and grow toward gravity. (Photo from Wilkins 1988, used with permission.)

Herbicides

The synthetic auxin 2,4-D (Weed-B-Gon®) is an inexpensive herbicide with relatively low mammalian toxicity that selectively kills broadleaf dicots (e.g., killing dandelions in a lawn without killing the grass; Figure 13.6). Grasses are able to inhibit the activity of the synthetic auxins used in these herbicides, making them less susceptible than dicots when such compounds are applied at recommended rates.

Prevention of leaf abscission

Growing leaves produce IAA, but stresses such as drought, nutrient deficiencies, or cold—including cool autumn temperatures—cause the production of IAA to decrease. A decrease in IAA production increases the sensitivity

Figure 13.5 Plants with strong apical dominance have a conical form (a), while plants with weak apical dominance have a more spherical shape (b).

of cells at the base of the petiole to ethylene, which can result in abscission. Synthetic auxins are sometimes sprayed on apple or orange trees to inhibit preharvest fruit drop.

Gibberellins

GA was discovered by plant pathologists in Japan studying rice (*Oryza sativa*) infected with the fungus *Gibberella fujikuroi*. These rice plants were pale and spindly, and did not produce seed. By 1939, the compound responsible for excess growth had been isolated from the fungus and named gibberellin A. It was subsequently discovered that gibberellic acids are also produced by plants. Although more than 130 different naturally produced gibberellins or gibberellic acids (GAs) have been isolated from plants, most are inactive precursors or degradation products of active GAs. Of the **gibberellins** found in plants, only a few are biologically active—GA_1 in many plants; GA_4 in others.

Location of GA synthesis

GAs are synthesized in immature seeds, in root and shoot apical meristems, and in developing anthers and pollen. GA is active in these tissues as well as in other tissues.

Figure 13.6 Dicots such as dandelions are more sensitive to auxin-related herbicides. This dandelion has been exposed to 2,4-D causing the stem to curl.

Transport of GA

Only auxins are transported in a polar fashion. GAs and other plant hormones (except ethylene, which is a gas) are transported in the xylem and phloem.

Effects of GA

Cell division and elongation

GA stimulates plant stem elongation by increasing cell division and subsequent elongation in the intercalary meristem at the base of the internode. GA can be applied to dwarf plants and cause them to attain normal height

Figure 13.7 The plant in the center is the maize inbred line B73, the dwarf plant on the left is a GA-insensitive near-isogenic line of B73, and the plant on the right is B73 treated with GA_3. (Courtesy of Isabel R. P. de Souza.)

or applied to normal plants to increase growth (Figure 13.7). Dwarf plants may be deficient in the production of biologically active GAs, or may be unresponsive to GAs. Some "response" mutants are tall because they lack normal feedback mechanisms that repress excessive GA synthesis. GA acts on cell elongation by increasing enzymes that break hydrogen bonds in the wall but not through increased H^+-ATPase activity as with auxin. GAs also increase the rate of cell division in intercalary meristems.

Flowering

Many biennial plants have short stem internodes in their first year (rosette form), then undergo internode elongation (**bolting**) and flowering in year two (Figure 13.8). An extended cold period will switch off genes that inhibit flowering. In long-day plants, long days increase GA synthesis, resulting in

Figure 13.8 A celery plant that has bolted following vernalization. (Courtesy of Dan Drost.)

stem elongation and flowering. The application of GA to the rosette form of these plants can substitute for cold temperatures and long days and induce bolting and, sometimes, flowering.

Inhibitors of GA

Inhibitors of GA are used to produce more-compact poinsettias, lilies, and chrysanthemums. For example, chrysanthemums are short-day plants that flower naturally in autumn but are often sold for Memorial Day in late spring. To promote flowering out of season, cuttings are exposed to extended daylengths while they develop vegetatively, the apical bud is removed to promote branching, and flowering is induced by exposure to short days. A compact form is desirable for potted chrysanthemums, so cuttings are treated with a GA inhibitor to reduce stem length and all but the terminal buds on each branch is removed, so the remaining flowers will be larger. The result

Figure 13.9 A chrysanthemum developed for sale in late spring (a) and a chrysanthemum blooming in a garden in autumn (b).

is a compact plant with a smaller number of large flowers (Figure 13.9a) in contrast to the many small flowers of a naturalized plant flowering in the autumn (Figure 13.9b).

Plant breeding to alter the effects of GA in wheat and rice were the basis for the Green Revolution for which Norman Borlaug was awarded the Nobel Peace Prize in 1978. Dwarf varieties of these cereal grains redirect photosynthate that would have been used for stems into seeds, increasing the "harvest index" or the proportion of the shoot biomass that is grain. Even when fertilized with high rates of nitrogen, dwarf wheat plants are less like to lodge (fall over), which would make them difficult to harvest.

Seed formation and germination

Genetically seedless grapes are treated commercially with GA to replace the GA that would have been produced by developing seeds. The added GA increases the length of branches in the grape cluster, as well as the size of fruits. In germinating seeds of the cereal grain barley, the cotyledon produces GA when germination begins (Figure 13.10). The GA is transported to the aleurone layer, a layer of living cells surrounding the nonliving endosperm that stores starch. The GA causes cells in the aleurone layer to produce the enzyme α-amylase, which is secreted into the endosperm and α-amylase begins the breakdown of amylose (starch). In commercial malting, GA is added to increase the rate of starch breakdown. Subsequently, β-amylase hydrolyses the short starch chains to glucose pairs (maltose), and the enzyme maltase hydrolyses these to glucose, which can be used by the germinating seedling for growth (or fermented by yeast to ethanol).

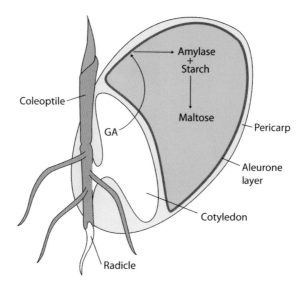

Figure 13.10 In germinating barley, GA from the cotyledon migrates to the aleurone layer, where it promotes the synthesis of α-amylase. Starch in the endosperm (blue) is hydrolyzed to maltose by α-amylase.

Cytokinins

The chemical structure of **cytokinins** is related to adenine, one of the four "bases" that are the key components of DNA (the other three bases are guanine, thymine, and cytosine). The naturally occurring cytokinin **zeatin** was isolated from corn (*Zea mays*) seed endosperm in 1963 and is the most common cytokinin in higher plants.

Location of cytokinin synthesis

In angiosperms (flowering plants), most cytokinin is synthesized in the root, transported to the shoot in the xylem stream, and promotes cell division in the shoot. Cytokinins are also synthesized in shoot apical meristems, axillary buds, and young leaves, and in some pathogens and insects that attack plants.

Effects of cytokinins

Cell division

Cytokinins interact with auxin to regulate the process of cell division. Cytokinins also promote the growth of axillary buds and interact with auxin to regulate apical dominance. Cytokinin overproduction due to pathogens can

Figure 13.11 Many witches' brooms are caused by the overproduction of cytokinins.

lead to the overstimulation of axillary bud growth and the production of a dense cluster of highly branched dwarf shoots known as a witches' broom (Figure 13.11).

Organogenesis

Organogenesis is the formation of new roots or shoots. Plant cells that have stopped dividing and that have differentiated can resume cell division, as in response to wounding. Sterilized segments of plant tissue can be induced to proliferate (form callus tissue) when grown on a medium with nutrients and a balance of auxin and cytokinin.

The subsequent differentiation of callus into root or shoot tissue is determined by the ratio of auxin to cytokinin. If the ratio of auxin to cytokinin is high, roots will form, and if the ratio is low, shoots will form. Varying these ratios allows the formation of complete plants. In the intact plant, high levels of nitrogen and other nutrients in the soil stimulate cytokinin production, which in turn stimulates shoot growth. In contrast, low nutrient levels result in low root cytokinin, which stimulates root growth and redirects plant resources to the acquisition of soil nutrients.

Delay of senescence

Senescence is the selective removal of nutrients and organic compounds from an organ, followed by death. Cytokinins are produced by young leaves and act to identify them as sinks for sugars and amino acids. Cytokinins applied to mature leaves act to delay senescence (Figure 13.12).

Figure 13.12 This photograph shows that cytokinins can delay senescence by mobilizing nutrients to leaves with higher cytokinin content. The stem of a 14-day-old kidney bean (*Phaseolus vulgaris*) was cut 4 in. below the unifoliolate leaves, and placed in a beaker of water. The two unifoliolate leaves were treated with benzyladenine (BA), a synthetic cytokinin, every fourth day. In control seedlings, the unifoliolate leaves would be senescent after 1 week. When treated with BA, the unifoliolate leaves remained green for more than 3 weeks, but the developing trifoliolate leaf senesced. (Figure from Leopold and Kawase 1964, used with permission.)

Ethylene

Ethylene is a gas that inhibits growth and promotes senescence. In the 1800s, it was observed that the trees near gas street lamps became defoliated while trees at a greater distance did not. Ethylene was subsequently identified as a by-product of the combustion of natural gas (methane), and was shown to cause defoliation and inhibition of plant growth. However, the fruit-ripening effects of some fruits on others provided the first clue that plants themselves produced ethylene. Ethylene is synthesized from methionine, a sulfur-containing amino acid, but ethylene has the simple formula $H_2C = CH_2$. The chemical intermediate between methionine and ethylene is ACC (1-aminocyclopropane-1-carboxylic acid), and ethylene synthesis is regulated by the synthesis and oxidation of ACC.

Effects of ethylene

Fruit ripening

An increase in ethylene in ripening fruits is followed by a climacteric increase in respiration rate (Figure 11.1). In ripening fleshy fruit, starch is converted to sugar, the chlorophyll in green fruits is degraded and other pigments such as anthocyanins and carotenoids are synthesized, and the walls of cells soften, caused by the production of the enzymes cellulase and pectinase, which break down cellulose and pectin, structural components of the cell wall (Figure 13.13).

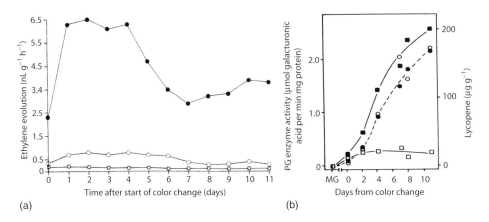

Figure 13.13 (a) The evolution of ethylene from normal ripening tomato (*Lycopersicon esculentum*) fruit (closed circles) or plants in which the last step of ethylene synthesis has been partially (open circles) or fully (open squares) blocked. (Figure from Hamilton et al. 1990, used with permission.) (b) The synthesis of polygalacturonase (PG), a pectinase, increases in normal ripening tomatoes (closed squares) along with the red pigment lycopene (squares), while PG synthesis has been blocked (open squares) in genetically transformed fruit. (Figure from Smith et al. 1988, used with permission.)

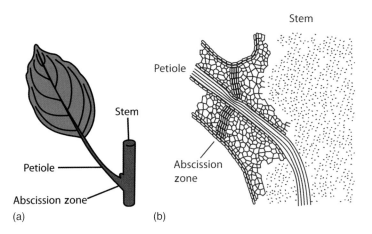

Figure 13.14 (a) The location of the abscission zone at the base of the petiole, and (b) the digestion of cell walls is followed by suberization of remaining cell walls within the abscission zone.

Fruits that respond to ethylene can be stored in low temperatures and low oxygen concentrations that inhibit ethylene biosynthesis. When stored fruits are needed for market, they are exposed to a small amount of ethylene. This causes fruits to experience a **climacteric**, the increase in respiration that results from increased metabolic activity.

Abscission

Healthy leaves produce auxin that moves through the petiole to the stem. The auxin production of leaves exposed to disease, drought or cold is reduced, which increases the sensitivity of cells in the **abscission zone** at the base of the petiole to ethylene (Figure 13.14a). As in ripening fruit, ethylene causes an increase in cellulase and other enzymes that degrade cell walls in abscission zone cells (Figure 13.14b), which is followed by leaf drop. Some fruits are sprayed with ethephon, which breaks down to ethylene, to increase the rate of ripening or to synchronize abscission and allow a shorter harvesting period.

Thigmomorphogenesis

Thigmomorphogenesis is the response of plants to mechanical disturbance, such as shaking (Figure 13.15). Plants that are physically disturbed produce more ethylene, and become stunted relative to unshaken plants. As in some other ethylene responses, changes in shoot growth are mediated through effects of ethylene on auxin and abscisic acid levels. Because of this ethylene response, natural mechanical disturbances such as wind are less destructive because the plant is shorter and sturdier.

Figure 13.15 Ethylene is produced when tomato (*Lycopersicon esculentum*) plants are shaken, causing stunting after 28 days. The plant on the left was undisturbed, the plant in the center was shaken for 30 seconds each day, and the plant on the right was shaken for 30 seconds twice each day. (Courtesy of Cary Mitchell.)

Seed germination

Ethylene plays a role in dicot seedling germinating of seeds planted in soil, causing the formation of a hook below the shoot apical meristem. Ethylene inhibits growth on the inside of the hook by creating an auxin gradient from the inner to the outer tissues. A hook below the shoot apical meristem allows the meristem to be dragged up to the soil surface, protecting the tenderest tissues. Once the hook reaches the soil surface, light is sensed by phytochrome which inhibits ethylene synthesis, causing the hook to straighten.

Response to waterlogging

The intercellular air spaces of the roots of waterlogged plants are filled with water, and therefore the root becomes starved for the oxygen needed for respiration. The accumulation of ethylene is elevated in roots of waterlogged plants, which results in the breakdown of cell walls within the root, creating large airspaces in the root cortex (Figure 13.16). These roots then resemble those of hydrophytes in which the formation of aerenchyma improves gas movement to cells.

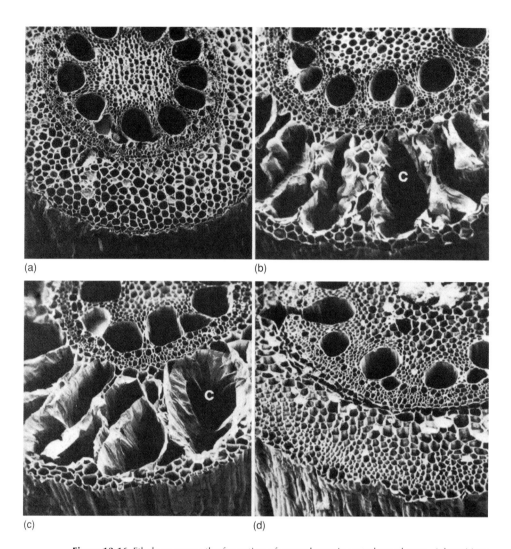

Figure 13.16 Ethylene causes the formation of aerenchyma in waterlogged roots. Adventitious roots of maize plants grown (a) in well-oxygenated hydroponic solution, (b) solution aerated with oxygen and ethylene, (c) nonaerated solution, and (d) solution aerated with nitrogen gas. Roots aerated with nitrogen (d) appear no more waterlogged than roots aerated with oxygen (a), while roots aerated with oxygen and ethylene (b) have developed aerenchyma (C, cortical air spaces), as have nonaerated roots (c). (Figure from Drew et al. 1979, used with permission.)

Abscisic acid

Abscisic acid (ABA) is synthesized in the plastids (chloroplasts or amyloplasts) of nearly all plant tissues, and is translocated in both the xylem and phloem. ABA is considered an inhibitor of growth, most notably in seeds. ABA was purified from extracts of dormant buds of woody plants and was therefore named "dormin." The same compound had been found to stimulate the abscission of cotton fruits, and had been named "abscisin." The name abscisic acid was adopted but ABA is no longer thought to control abscission.

Effects of ABA

Stomatal closure

ABA synthesis is increased in plants under drought stress. Drought-stressed leaves make large amounts of ABA, and roots in dry soil also synthesize ABA that can be transported in the xylem stream to the shoot. ABA in leaves acts at the guard cell plasma membrane, causing anions (Cl^- and malate) to leave guard cells; the anions are followed by potassium (K^+). The guard cells become flaccid because water follows K^+ out of guard cells by osmosis, causing them to collapse over the stomatal pore (Figure 8.13). In roots, elevated ABA under water stress acts to *increase* root elongation.

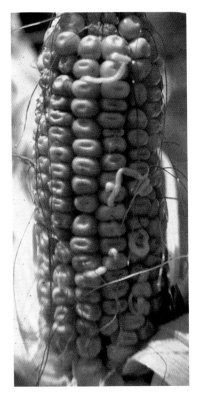

Figure 13.17 Vivipary (precocious germination) in kernels of a low-ABA mutant of maize. (Courtesy of David Hole.)

Seed dormancy

During seed development, an embryo is formed that eventually becomes the plant body. The endosperm is formed at the same time, and storage in the endosperm occurs following cessation of cell division in the embryo (Figure 13.10). Following storage of the food resources that will be needed during germination, seeds become dehydrated. As seeds accumulate food reserves, relatively high levels of ABA inhibit growth of the embryo into a plant. Over time, ABA levels decrease in mature seeds, and seeds are then able to germinate. ABA acts by opposing the action of GA, which stimulates the amylase activity needed for seed germination. GA stimulates gene expression of α-amylase, and ABA represses expression of the same gene. When ABA is deliberately reduced or is unusually low, or tissue is ABA insensitive due to a mutation, the result is **vivipary**, or the germination of seeds on the seedhead (Figure 13.17).

Chapter 14
Secondary plant products

In Chapter 12, we saw how plants read and adapt to the seasonal changes in their environments by adjusting their growth, reproduction, and chemical makeup. In Chapter 13, we saw how plant hormones are used to adjust to less predictable short-term changes such as the loss of an apical meristem or drought stress. In this chapter, we discuss the use of secondary metabolites to mount defenses against attack or compete for the attention of a pollinator. Many of these compounds have also proven useful for human purposes, some since prehistoric times.

The term **"secondary plant products"** or "plant secondary metabolites" is applied to plant products that have no apparent role in the fundamental metabolism of the growth and development of the plant. Many are important as toxins or feeding deterrents with respect to herbivores (animals that feed on plants), and therefore to the survival of the plant.

Types of secondary plant products

There are three major chemical classes of secondary metabolites:

1. Terpenes (insoluble in water like lipids)
2. Phenolics (derived from carbohydrates)
3. Alkaloids (derived from amino acids)

Terpenes

Terpenes are produced by many plant species, most notably in the resin of conifers. The name "terpene" is derived from "turpentine," a volatile component of resins. **Terpenes**, the largest group of secondary metabolites, are built from 5-carbon isoprene units and are grouped by the number of subunits they

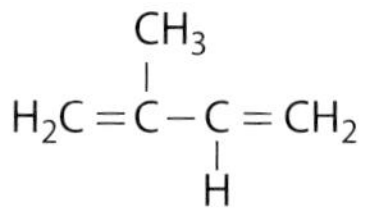

$$H_2C = \underset{H}{\overset{CH_3}{\underset{|}{C}}} - C = CH_2$$

Isoprene unit

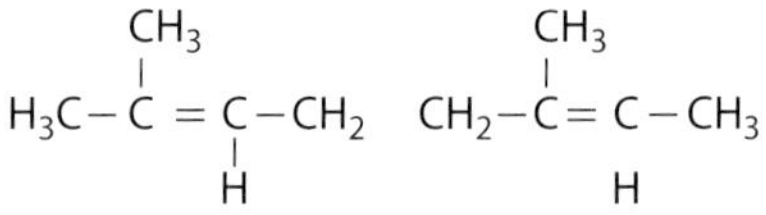

$$H_3C - \underset{H}{\overset{CH_3}{\underset{|}{C}}} = C - CH_2 \quad CH_2 - \underset{H}{\overset{CH_3}{\underset{|}{C}}} = C - CH_3$$

Monoterpene

$$H_3C - \underset{H}{\overset{CH_3}{\underset{|}{C}}} = C - CH_2 - \left(CH_2 - \underset{H}{\overset{CH_3}{\underset{|}{C}}} = C - CH_2\right)_n - CH_2 - \underset{H}{\overset{CH_3}{\underset{|}{C}}} = C - CH_3$$

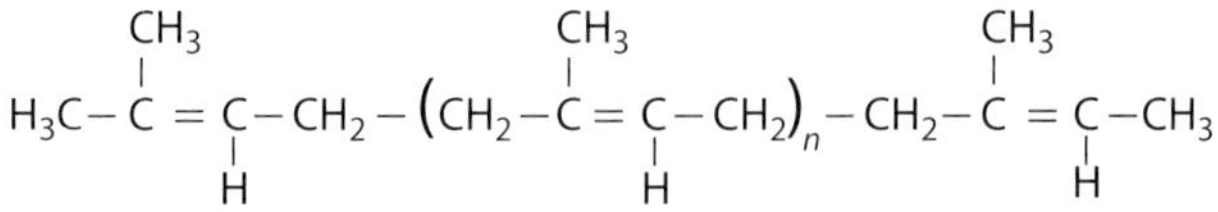

Polyterpene

Figure 14.1 The building block of terpenes is the 5-carbon isoprene unit. A monoterpene consists of two isoprene units and has 10 carbons, and a sesquiterpene (one and one-half monoterpenes) has 15 carbons. Larger and more complex compounds are also built from isoprene units.

contain: monoterpenes contain two, sesquiterpenes three, diterpenes four, and so on (Figure 14.1). Some terpenes are part of primary metabolism: gibberellic acid and abscisic acid are plant growth regulators, and the carotenoid pigments and tail that anchors chlorophyll in the thylakoid membrane are terpenes.

Monoterpenes

Essential oils contain volatile monoterpenes that are stored in glandular hairs of the epidermis in a modified cell wall space (Figure 14.2). They are used in perfumes and food flavorings. Examples of food plants whose essential oils are the basis for their use are peppermint, lemon, basil, oregano, thyme, and sage. Pyrethrins from the seed cases of *Chrysanthemum cinerariafolium* are insecticidal monoterpenes, and are desirable pesticides because they have low environmental persistence and low mammalian toxicity (or a high LD_{50}, the dose that is lethal for half a population).

Diterpenes

Resins contain mono- and diterpenes and are formed or stored in resin ducts (Figure 14.3). When resin ducts are pierced, the outflow of resin physically blocks feeding, and the terpenes serve as a toxic chemical deterrent to feeding. Resins polymerize (the diterpenes become linked to each other) on exposure to air and seal the wound.

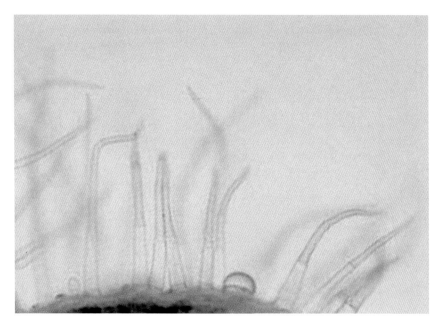

Figure 14.2 Glandular hairs from a leaf of catmint (*Nepeta*).

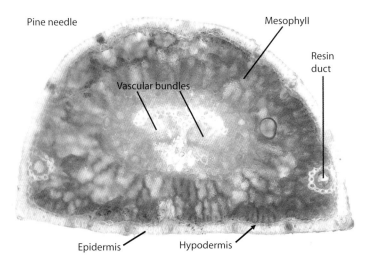

Figure 14.3 Cross section of a pine needle showing the location of a resin duct.

Terpenes Used by Both Sides of a Predator–Prey Relationship

Mountain pine bark beetles (*Dendroctonus ponderosae*) have devastated forests throughout the Rocky Mountains during an outbreak that started in 1996. Forests are susceptible because they have been planted as monocultures, wildfires have been suppressed, and trees have been stressed by prolonged drought during this period. Mountain pine beetle adults emerge in late summer, fly to new trees, bore into the trees, and lay eggs in the phloem, the vascular cambium, and the sapwood. The eggs hatch into larvae that live under the bark until the next summer, feeding on the phloem and effectively girdling the tree (Figure 14.4).

Female beetles, acting as "pioneers," scout a tree and release sex and aggregation pheromones (scent signals) that attract males and other females that swarm onto the tree, often overwhelming its defenses. The beetles carry along pathogenic, blue-staining fungal partners in pouches on their bodies that spread into the sapwood, reducing the flow of resin and killing the trees attacked by the beetles. The dead trees—now including much of the lodgepole pine population—drive away tourists and add to the wildfire hazard. Beetles are killed by 1 or 2 weeks of temperatures close to $-40°C$ ($-40°F$) and winter temperatures have not been dropping low enough to control the insects.

The initial defense mounted by conifers as the pioneer beetles bore through the bark is an outflow of resin that drowns the beetles, encasing them in resin, flushing out the wound, and when the turpentine evaporates, leaving a physical barrier to seal the wound (Figure 14.5). The resin contains turpentine and rosin composed of mono-, sesqui-, and diterpenes (10-, 15-, and 20-carbons, respectively). The volatile monoterpenes in the resin have a complicated role in the interaction between

Figure 14.4 Resin has flowed from ducts and hardened on the surface of the trunk of this conifer. (Courtesy of James Sinclair.)

Figure 14.5 A gallery of the mountain pine beetle showing the vertical groove into which eggs were laid and larvae that have migrated to each side. The blue cast of the wood is caused by a pathogenic fungus that is introduced by the beetle. (Courtesy of Ingrid Aguayo, Colorado State Forest Service.)

predator beetle and host tree, allowing the beetle to recognize its host, contributing to the synthesis of the pheromones that attract the beetle swarm, and signaling an SOS to predators and parasitoids of the beetles. Trees that have been damaged by lightening or just physiologically compromised by drought emit a plume of monoterpenes that attract the pioneer beetles (Trapp and Croteau 2001). In some species of bark beetles, most interestingly, the aggregation pheromones that attract the swarm of beetles are actually synthesized from the volatile monoterpenes in the resin. As a last resort defense, the same terpenes that provide host recognition for bark beetles and the pheromones that pioneer beetles use to signal a swarm also attract predators and parasitoids that attack the bark beetles.

Not all resins are toxic. The first popular chewing gum was made from chicle, a resin from the sapodilla tree. Thomas Adams, an American inventor, attempted unsuccessfully to use chicle to make rubber (also synthesized from terpenes). The sapodilla tree is native to Mexico and Central America, and the resin had been chewed since the time of the Mayans. Abandoning his effort to synthesize rubber, Adams added sugar, flavorings, and a candy shell and created a popular treat.

Triterpenes

Milkweed (*Asclepias*) contains triterpenes with a sugar side chain (glycoside). The larvae (caterpillar) of monarch butterflies feed on milkweed and can accumulate the toxin in their bodies. When monarchs metamorphose into butterflies, the terpene is stored in its wings (Figure 14.6) making them

Figure 14.6 The monarch butterfly ingests terpene glycosides from leaves of milkweed as a caterpillar. As a butterfly, the toxin is stored in the wings, making the butterfly poisonous and giving it a foul taste.

toxic to predators such as birds. The viceroy butterfly mimics the appearance of the monarch and reaps the benefit of the toxin without having to consume it.

Digitalin, a triterpene from foxglove (*Digitalis purpurea*), slows and strengthens the heart muscle. It was first developed as a drug in the late 1700s from a folk remedy, a tea made from the leaves of purple foxglove that was used to treat the condition known then as dropsy. William Withering's 1785 publication of the relationship between dose of foxglove and response in patients is considered the beginning of modern pharmacology. Digitalin was shown to be an effective drug long before the cause of the condition, congestive heart failure, was known.

The neem (*Azadirachta indica*) tree is a highly drought-tolerant native of southern Asia. Oil extracted from neem seed kernels contains a triterpene that is effective as a feeding deterrent at very low concentrations (parts per billion), and is used on plants as an organic insecticide. It does not kill insects that feed on treated foliage, but inhibits their ability to feed, breed, or metamorphose. It has no apparent mammalian toxicity and has many therapeutic uses in traditional medicine.

Tetraterpenes

The carotenoids—yellow, orange, and red photosynthetic pigments such as carotene and lycopene—are tetraterpenes, and the plant growth regulator abscisic acid (ABA) is derived from a carotenoid precursor.

Polyterpenes

Rubber is a polyterpene, made of as many as 6,000 terpene units in linear chains. Vulcanization is the cross-linking of these chains via added sulfur, and prevents rubber from cracking in cold or becoming sticky in heat. Charles Goodyear discovered vulcanization by heating a mixture of sulfur and rubber, accidentally by some accounts. Rubber occurs as small particles suspended in a fluid (**latex**), which is found in laticifers, channels located in the inner bark. Latex from the rubber tree (*Hevea brasilliensis*) is 30–50% rubber; the rest is mostly water. To tap the rubber tree, a diagonal cut is made in the bark without injuring the vascular cambium. Another vertical cut guides the rubber into a collection vessel. Every other day a thin layer of bark is shaved from the top of the cut to open clogged laticifers and maintain flow. This process continues down one side of the tree, then the other side of the tree is tapped and the first side is allowed to heal. Rubber balls have been found in Central and South America that date to as long ago as 1600 BC. The English term "rubber" was applied when it was discovered that this substance could rub pencil marks off paper.

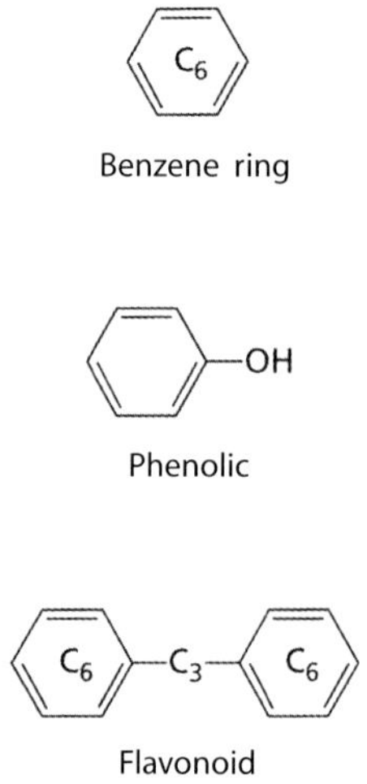

Figure 14.7 The fundamental form of a phenolic is a benzene ring plus a hydroxyl group. Flavonoids are synthesized from two phenyl rings.

Phenolics

Most **phenolics** are derived from the amino acid phenylalanine. The shikimic acid pathway is the primary pathway used by plants to convert carbohydrates to the three aromatic (containing a 6-carbon benzene ring) amino acids: phenylalanine, tyrosine, and tryptophane. Animals do not possess the shikimic acid pathway and therefore cannot synthesize these three amino acids. Glyphosate (Roundup® herbicide) kills plants by blocking a step in the shikimic acid pathway.

Phenolic compounds contain a phenol group, which is a 6-carbon aromatic (benzene) ring plus a hydroxyl (−OH) group (Figure 14.7). Lignin is a large molecule composed of phenolic monomers. Flavonoids have a basic structure of two rings linked by three carbons, and **tannins** are polymers of flavonoid units.

Allelopathy

Allelopathy is the effect that a chemical compound from one plant can have on another plant sharing its environment. It is often but not necessarily a negative effect. Simple phenolic compounds have been shown to inhibit the germination and growth of many plants, and are thought to be allelopathic in some cases. Alfalfa (Figure 14.8) is auto (self) toxic due to allelopathy, and results in the death of germinating alfalfa seedlings. A field of alfalfa usually cannot be replaced by a new seeding of alfalfa, so some other crop must be grown in the field first. These allelopathic phenolics are water-soluble and therefore can sometimes be washed out of the root zone by irrigation before reseeding alfalfa into alfalfa.

Figure 14.8 Established alfalfa plants reduce competition by exuding phenolics that inhibit root growth of alfalfa seedlings.

Lignin

Lignin is a phenolic compound, and is the most abundant organic substance in plants after cellulose. Formation of lignin occurs in the thickened walls of xylem and fiber cells. Lignin formation displaces water and seals the cell wall to allow water movement under tension through the xylem. It provides rigidity and strength, and is considered to be the main factor that allowed plants to colonize dry land, which required the transport of water from roots to shoots. Lignin is a physical deterrent to feeding and is also synthesized

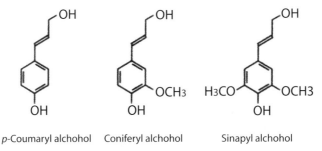

p-Coumaryl alchohol   Coniferyl alchohol   Sinapyl alchohol

Figure 14.9 Lignin is built of phenolic monomers that are secreted into the cell wall. Bonds among these monomers can form at several sites on each molecule and form in three dimensions, making lignin difficult to digest. (Monolignol structures courtesy of Ron Hatfield.)

in response to infection or wounding. It coats cellulose and other cell wall polymers, reducing their digestibility. Lignin is synthesized from phenolic monomers (Figure 14.9) in the cell wall through the formation of chemical bonds between a variety of sites on each monomer. Because many enzymes would be required to break such a diverse array of bonds, lignin is minimally digestible. There is interest in using genetic engineering to modify the way plants make lignin to make it easier to remove lignin from wood pulp, and to increase the digestibility of the fiber cells in forage species.

Flavonoids

Flavonoids are phenolic compounds that are employed by plants as the visual and olfactory attractants needed for pollination and seed dispersal. Anthocyanins are flavonoids that are the source of most red, pink, purple, and blue colors in flowers and fruits (Figure 14.10). Anthocyanins are water-soluble and therefore are stored in the vacuole (Figure 14.11), while the carotenoids that provide the yellow, orange, and red plant pigments are insoluble in water and stored in plastids.

In autumn, cool temperatures and decreasing daylengths signal the leaves of woody species to senesce. Senescence can be seen as a decrease in green color (chlorophyll), an increase in yellow color (carotenoids), and in some leaves, an increase in red color (anthocyanins) (Figure 14.12). During senescence, carbohydrates are plentiful in leaves, but it is critical for trees—which often exist in low-nitrogen environments—to recapture the nitrogen in proteins in the thylakoid membranes and stroma of chloroplasts. Leaves that turn red in the fall, such as burning bush (*Euonymus*), synthesize anthocyanins from leaf carbohydrates and store them in vacuoles (Figure 14.13).

(a) (b)

Figure 14.10 Cornflower (a) and begonia (b) are examples of the blue and pink flowers colored by anthocyanins.

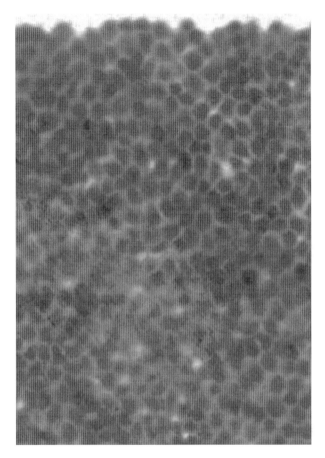

Figure 14.11 The vacuoles of the cells of this rose petal are filled with anthocyanins.

Then chlorophyll is unbound from chloroplast proteins, unmasking the yellow carotenoid photosynthetic pigments. Anthocyanins function to absorb the light that would otherwise damage cellular components through photooxidation and interfere with the recovery of cellular nutrients (Feild *et al.* 2001). Anthocyanins are also osmotically active and therefore protect senescing cells against cold and drought.

The epidermis of the leaves of all plants contains flavonoids that protect against UV-B radiation. These compounds absorb light in the range 280–320 nm but allow visible light to pass through uninterrupted for photosynthesis. The patterns formed on flowers by UV absorbance by flavonoids can form nectar and pollen guides (Figures 6.18 and 14.14). Flavonoids are also used by legumes to attract nitrogen-fixing rhizobia bacteria from the soil to the root.

Tannins

Tannins are larger molecules composed of polymerized flavonoids. Tannins that bind to the collagen proteins in animal skins prevent microbial attack

246 Structure and Function of Plants

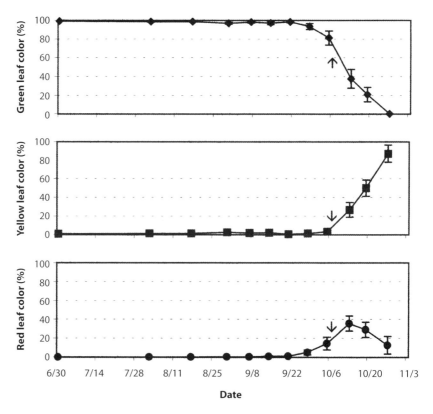

Figure 14.12 Sugar maple (*Acer saccharum*) leaf color from early summer through mid-autumn. Initial color change was an increase in red on September 28, and an increase in yellow and a decrease in green were both detected on October 13 following the first frost on October 6 (arrows). The decline in red in mid-October was due to early senescence of red leaves. (Figure from Schaberg *et al.* 2003, used with permission.)

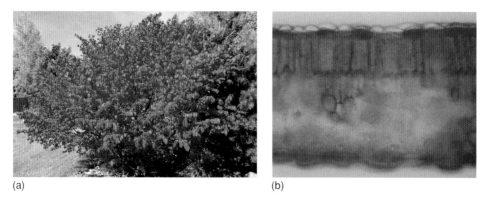

Figure 14.13 (a) The bright autumn color of burning bush (*Euonymus*) is the result of accumulation of anthocyanins in the vacuoles of epidermal cells, as seen in the cross section of a leaf blade (b).

Figure 14.14 Common evening primrose (*Oenothera biennis*) photographed in visible (a) or in ultraviolet light (b). (Photographs © Bjørn Rørslett, www.naturfotograf.com.)

and preserve the skins, and are therefore used in making leather. There is clear evidence that Egyptians used tanned leather to make sandals more than 3,000 years ago. The barks of oak, spruce, and chestnut trees are all commercial sources of tannins; a high level of tannins in nonliving heartwood prevents microbial decay.

Tannins are considered to be toxins that reduce the growth and survival of herbivores. They are feeding repellents that bind salivary proteins, and plant parts high in tannins are usually avoided by cattle and deer. Apples, blackberries, tea, and red wine contain tannins, which provide a desirable level of astringency (a dry or puckery sensation in the mouth caused by binding to salivary proteins). Leaves accumulate tannins in sacs that separate

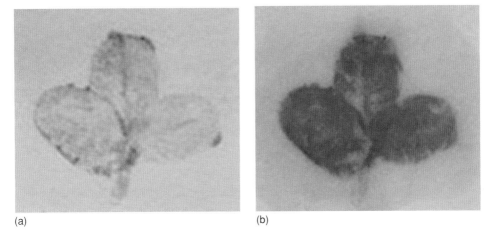

Figure 14.15 The sap pressed from leaflets of tannin-containing sainfoin (*Onobrychis viciifolia*) (a) and stained for tannins with vanillin-HCl (b).

them from cellular proteins such as rubisco, but when leaves are crushed or chewed, herbivores are exposed to the tannins (Figure 14.15). Rodents and rabbits cope with tannins by producing salivary proteins that have a high affinity for tannins, diminishing their toxic effect. Plant tannins bind proteins in the digestive systems of herbivores, which can have a negative effect on herbivore nutrition. However, there can be a beneficial effect of moderate tannin levels on protein utilization of ruminant livestock. Forages with a high soluble protein content such as alfalfa and the true clovers can cause bloat, a condition in which gas accumulates in the rumen and which can kill ruminants in as little as 20 minutes. Bloat is prevented by tannins binding to plant proteins in the rumen. Tannins can also have a direct negative effect on some internal parasites of ruminants.

Alkaloids

Alkaloids are nitrogen-containing secondary metabolites that have potent pharmacological effects in vertebrate animals. Most alkaloids are synthesized from amino acids. The nitrogen atom in alkaloids is usually part of a "heterocyclic" ring, or a ring containing nitrogen in addition to carbon (Figure 14.16). Alkaloids are so named because the nitrogen group is positively charged at the pH of plant cells, making the molecule basic or alkaline. Some alkaloids are structurally similar to neurotransmitters causing them to have a potent effect on the central nervous system.

Nicotine

Nicotine is the bioactive component of plants in the nightshade family (Solanaceae), particularly tobacco (*Nicotiana tabacum*). In the plant, nicotine is synthesized in the root and accumulates in leaves. It is a neurotoxin, particularly toward insects and has historically been used as an insecticide. In low concentrations in mammals nicotine is a stimulant, causing the release of adrenaline. Nicotine inhaled through the lungs crosses the blood–brain barrier in as little as 7 seconds.

Caffeine

Coffee (*Coffea*), tea (*Camellia*), and chocolate (*Theobroma cacao*) all contain caffeine, an alkaloid that is a central nervous system stimulant and mild diuretic. Caffeine is absorbed from the stomach and intestines. Caffeine mimics the feeling produced by the hormone adrenaline and makes analgesics (pain relievers that do not decrease consciousness) significantly more effective. Caffeine has been consumed by humans chewing the seeds, leaves, or bark of plants since the Stone Age. In plants, caffeine acts as an insecticide, paralyzing and killing some insects, and as an allelopathic agent, reducing competition by inhibiting the germination of seeds. A single shot of

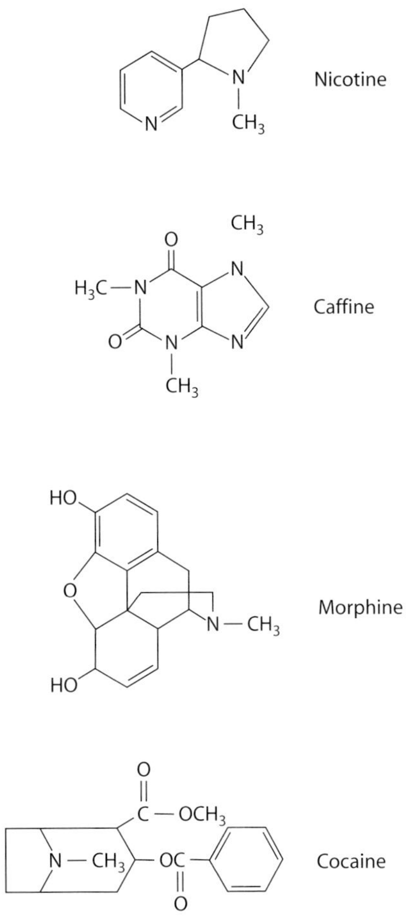

Figure 14.16 Structures of well-known alkaloids showing the heterocyclic rings common to this class of compounds.

espresso contains about 40 mg caffeine, while a cup (250 mL) of drip-brewed coffee contains about 100 mg. Tea contains about half the caffeine of coffee. The discovery of tea is credited to a Chinese emperor in about 3000 BC, and the discovery of coffee to the observations of a shepherd in Ethiopia in the ninth century whose flock consumed the seeds of the coffee tree and was unable to sleep. Cacao is known to have been consumed by the Mayans as long ago as 600 BC.

Opium

Opium, morphine, and codeine are all alkaloids found in the latex of the opium poppy (*Papaver somniferum*). The seeds of this poppy contain very little alkaloid and can be used in cooking; the alkaloids are in the latex of the poppy fruit (Figure 14.17). The latex is harvested by slashing the capsules just

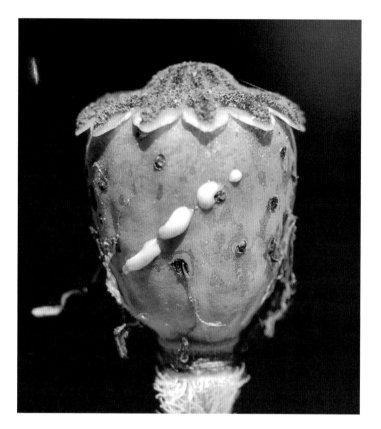

Figure 14.17 Latex from an immature poppy (*Papaver*) capsule.

deeply enough to cut into the latex vessels. The latex oozes out, is allowed to cure for a day, and is collected and dried. The same capsule can be tapped a number of times by a skilled worker.

There is evidence that opium poppy has been used by humans since the Neolithic period. Opium was used for surgery in the Islamic Empire between AD 600 and 1500 but became taboo in Europe during the Inquisition (AD 1300–1500). The British maintained a lucrative business selling opium to the Chinese even though the Chinese government had prohibited its use in 1729. In an attempt to stem a trade imbalance, Britain fought two Opium Wars with China, from 1839 to 1842 and from 1856 to 1860, forcing the Chinese government to tolerate the opium trade and yield the port of Hong Kong to Britain. The use of opium in China was not controlled until the establishment of the People's Republic of China in 1949.

Morphine
More than 25 alkaloids can be extracted from opium poppy latex, but the most abundant and the most potent alkaloid is the painkiller morphine. Because it is purified, an exact dose can be administered with a predictable

outcome, eliminating a problem in the use of opium. The narcotic alkaloids bind to receptors in the brain, spinal cord, intestines, and stomach. The dangers of morphine addiction were not fully appreciated until tens of thousands of Civil War soldiers returned addicted to the painkiller.

Herion
Heroin was synthesized from morphine in 1874, and was marketed as a nonaddicting substitute for morphine. Heroine is slightly different structurally from morphine, and the difference causes it to act more rapidly. In the nervous system, heroine is metabolized to morphine. The fortunes of the pharmaceutical companies Merck and Bayer were enhanced by the legal purification and marketing of morphine and heroine, respectively, although Bayer's most notable discovery is aspirin.

Papaverine and codeine
The third major alkaloid from the opium poppy, papaverine, does not affect the central nervous system. Paragoric is an effective treatment for diarrhea because the papaverine it contains controls internal muscle spasms. Codeine is another alkaloid from opium poppy latex. It is the most widely used opiate in the world because it is about 1/10th as potent as morphine and is therefore relatively safe. It is used as a pain reliever and in prescription cough syrups.

Cocaine

Cocaine, an alkaloid from the coca (*Erythroxylum coca*) plant, is a central nervous system stimulant. Coca has been used for thousands of years by the Andean natives of South America who chew the leaves. The purification of cocaine was the subject of a PhD dissertation in Germany in 1860. Subsequently, an Italian chemist Angelo Mariani developed a popular wine with coca leaves added called Vin Mariani. The ethanol in the wine solublized the cocaine from the coca leaves. Coca-Cola® was developed in Georgia as an American coca wine, but was transformed to a nonalcoholic patent medicine in response to local prohibition laws. It originally contained cocaine from coca leaves and caffeine from kola (*Cola nitida*) nuts. From 1904 to the present, coca leaves from which the cocaine has been extracted have been used to flavor the syrup from which Coca-Cola is made.

Poisonous wild plants

Ruminant livestock grazing rangeland in the Great Basin and Colorado Plateau are frequently poisoned by plants that contain alkaloids (Pfister *et al.* 2001). These plants can be plentiful on rangeland, and they are often beautiful plants that add nitrogen to the ecosystem that can be used by associated plant species. It is rarely economical to remove these plants using herbicides, so managing the livestock to avoid the plants when they are most toxic is often the best approach.

Figure 14.18 A high mountain meadow containing tall dark blue larkspur and shorter light blue lupine blooming among other wildflowers in mid-summer.

Figure 14.19 A stand of false hellebore in flower. (Courtesy of James Sinclair.)

Both larkspur (*Delphinium*) (Figure 14.18) and monkshood (*Aconitum*) contain toxic diterpenoid alkaloids which can cause paralysis and death in cattle from respiratory failure; sheep are less susceptible than cattle. Toxins are highest when vegetation is immature, so cattle prefer to graze larkspur from the time it flowers until plants form pods. Therefore, cattle can be allowed to graze rangeland infested with larkspur before the plants flower—the cattle will avoid the plants at this time—or after seeds shatter (Pfister *et al.* 2001).

Wild lupines (Figure 14.18)—there are more than 25 *Lupinus* species—can contain both quinolizidine and piperidine alkaloids. These compounds are highest in immature plants and seedpods, and can cause respiratory failure as well as birth defects in cattle. Since the developing calf is most susceptible during its first 100 days, avoiding rangeland with high lupine populations when cattle are in their first trimester of gestation is good management practice.

False hellebore (*Veratrum*; Figure 14.19) contains the alkaloid cyclopamine which causes birth defects, and sheep are especially susceptible because they readily consume the plant while cattle avoid it. False hellebore grows in dense stands in wet meadows that are common in early spring at high elevations. Since the window for fetal toxicity in sheep is short (14–33 days gestation), sheep should not be allowed into areas with stands of false hellebore when they are most vulnerable.

Glossary

Abscisic acid   A growth-inhibiting plant hormone that is active in the control of stomata during drought and was originally associated with dormancy and leaf fall

Abscission zone   A zone of cells at the base of a leaf petiole that is degraded with senescence, causing leaf drop

Action spectrum for photosynthesis   The relative rates of photosynthesis within the visible light spectrum

Active transport   The accumulation of molecules, such as sucrose or potassium, in cells against a concentration gradient through the expenditure of energy

Adventitious roots   Roots that develop from organs other than roots

Aerenchyma   Tissue containing large continuous internal air spaces for gas exchange

Aggregate fruit   A collection of small fruits borne on a single receptacle and derived from several ovaries within a single flower (e.g., raspberry)

Alkaloids   A group of nitrogen-containing secondary compounds with powerful physiological effects

Allelopathy   An effect of a chemical compound produced by one plant on another plant

Amino acids   Building blocks of proteins consisting of a central carbon, an amino group, a carboxyl group, and unique side group ("R")

Amyloplasts   Plastids that synthesize and store starch

Anion   A negatively charged ion, such as nitrate (NO_3^-)

Annual rings   Seen in the cross section of a tree trunk, concentric rings apparent because of the decrease in xylem element diameter from spring to autumn

Antenna complex   Aggregates of photosynthetic pigments on the thylakoid membranes that funnel light energy to a reaction center chlorophyll *a* (chl *a*) molecule

Anther   The structure within a stamen where pollen is formed

Apical dominance The suppression of the growth of axillary buds by the apical meristem, especially on the upper stem

Apical meristem The region of cell division that contributes to increase in the length of a stem or root

Apoplast The nonliving space outside the protoplast, including cell walls, through which water and dissolved substances can travel among cells

Aquaporins Protein channels in the plasma membrane specifically for the rapid movement of water

ATP (adenosine triphosphate) The energy-rich molecule synthesized by ATP synthase from ADP and protons (H^+) sequestered by reactions of the electron transport chain

ATP synthase The protein complex that includes a channel through which protons (H^+) flow from the lumen of the thylakoid membranes or the inner membrane space of mitochondria, forming ATP from ADP and inorganic phosphate

Auxin A hormone that promotes growth through cell expansion, inhibition of leaf senescence, apical dominance, and in response to stimuli such as gravity

Axillary bud A meristem that is the site of branch or flower formation, located at the junction of the stem and a leaf

Bacteroids The nitrogen-fixing form of rhizobia bacteria within legume root nodules

Bark The tissues of a tree trunk outside the vascular cambium, including both the phloem and the cork

Biochemical pathway A series of enzymatic reactions in which the product of one reaction is the substrate of the succeeding reaction

Biochemical (dark) reactions of photosynthesis The reactions in which energy from ATP and NADPH is used to add carbon dioxide (CO_2) to ribulose bisphosphate (RuBP)

Bolting The rapid growth in the length of a stem prior to flowering, often seen in biennial plants that form a rosette in their first year

Bud scales Modified leaves that protect apical and axillary meristems of woody species from freezing and dehydration in winter

Bulbs Fleshy leaves on an unelongated stem; used for vegetative propagation

Bulliform cells Large cells of the upper epidermis of grasses that function in leaf rolling during drought

C_3 photosynthesis The biochemical reactions in which CO_2 is added to the 5-carbon sugar RuBP, resulting in the formation of two molecules of the 3-carbon 3-PGA

C_4 photosynthesis Photosynthesis in which the initial capture of CO_2 from the atmosphere occurs in mesophyll cells, resulting in 14-carbon compounds that are shuttled to bundle sheath cells where CO_2 is released and used in C_3 photosynthesis

Calvin cycle The series of reactions central to C_3 photosynthesis in which CO_2 is captured and RuBP is regenerated

Calyx A collective term for the whorl of sepals that functions to protect the developing flower bud

Capillary water After precipitation, the water loosely held against gravity in small soil pores by surface tension

Carbohydrates Sugars such as glucose and fructose or polymers of sugars such as starch and cellulose

Carotenoids A group of lipid-like pigments that absorb in the green and red light wavelengths, and act as accessory pigments to chlorophyll in photosynthesis

Casparian strip A band of suberin and lignin in the inner primary wall of endodermal cells that allows roots to regulate the uptake of dissolved mineral nutrients

Catalyst An agent that promotes change without itself being changed

Cation A positively charged ion, such as potassium (K^+)

Cation exchange capacity (CEC) The relative abundance of nutrient cations held by ionic attractions to a soil; a measure of soil fertility

Cell division zone (root) The meristem located at the tip of each root that produces root tissue

Cellulose A structural carbohydrate composed of long chains of glucose molecules, which comprise the bulk of cell walls; the molecules are so large that they are synthesized at the plasma membrane and extruded into the cell wall space

Central mother cells The meristem initials in apical meristems of gymnosperms

Chelated nutrient An inorganic plant nutrient combined with an organic molecule for absorption by plants

Chlorophyll The major pigment of photosynthesis, which reflects green light and absorbs both blue and red light

Chloroplast The cellular organelle in which photosynthesis occurs

Chlorosis Reduced chlorophyll content due to a nutrient deficiency or waterlogging that results in yellowing of leaves

Clay The smallest soil mineral particles, less than 0.002 mm in diameter

Climacteric A rise in metabolic activity, and therefore respiration, during the ripening of some fruits that is caused by the hormone ethylene

Cohesion The strong attraction of water molecules for each other caused by their internal charge polarity

Collenchyma Support tissue that has thickened but nonlignified primary cell walls

Companion cells Phloem cells containing a nucleus that assume many of the metabolic functions of the associated sieve element and arising from the same mother cell

Complete flower A flower that has sepals, petals, stamens, and a pistil

Composite head An inflorescence in which a large number of small fertile (disk) flowers are attached to a flattened receptacle at the top of the peduncle, surrounded by sterile ray flowers with large petals

Compound leaf A leaf that consists of more than one leaflet

Contractile roots Roots that contract to pull the base of the shoot underground

Cork Protective secondary tissues that replace the epidermis of woody stems and roots; cork cell walls contain suberin that makes them waterproof and airtight

Cork cambium A lateral meristem that forms cork, the outer layer of a tree's bark

Corm Enlarged unelongated internodes at the base of the previous year's stem that are the source of carbohydrates to support the next year's growth

Corolla A collective term for all the petals of a flower, which may be composed of more than one layer and which often function to attract pollinators

Cortex A tissue composed of several layers of thin-walled cells inside the epidermis of the stem or root

Covalent bond A bond due to the sharing of one or more electrons by two atoms

Crassulacean acid metabolism (CAM) photosynthesis A system of temporal separation of the steps of photosynthesis in which CO_2 is absorbed at night, stored as malic acid in the vacuole, and released for use in C_3 photosynthesis during the day when stomata are closed

Critical photoperiod   The photoperiod required for flowering: a minimum daylength for long-day plants or a maximum daylength for short-day plants

Crop coefficient   A unitless value used to describe the relative amount of soil water that will be transpired by a particular crop

Cross-pollination   The transfer of pollen from a flower of one plant of a species to a flower of another plant of the same species

Cuticle   A layer of waxy cutin superimposed on the surface of epidermal cells of the shoot that reduces evaporative water loss

Cytokinin   A plant hormone synthesized in root tips that promotes cell division and inhibits senescence (aging) by attracting and retaining nutrients

Cytoplasm   The plasma membrane and contents of the cell, excluding the nucleus

Cytoplasmic membrane   (see **Plasma membrane**)

Cytosol   The liquid matrix of a cell, containing dissolved proteins, nutrient ions, and sugars, which bathes cellular organelles

Daylength   The number of hours of sunlight; daylength may change significantly with season depending on latitude

Day-neutral   Plants that flower at a given stage of maturity not affected by daylength

Dehydration reaction   An addition to a molecule with the loss of an oxygen and two hydrogens

Differentially permeable membrane   A membrane through which water can move, but that does not allow the movement of other molecules such as sugars

Diffusion   Movement of molecules such as salts in solution from an area of high concentration to an area of low concentration

Dioecious plant   A plant with staminate (male) and pistillate (female) flowers on separate plants (e.g., buffalo grass)

Dormancy   A state in which growth or development is suspended

Egg   The female gamete (reproductive cell)

Electron transport chain of photosynthesis   A series of oxidation–reduction reactions on the thylakoid membranes that result in the accumulation of hydrogen ions in the lumen of the thylakoid membranes and the synthesis of NADPH

Electron transport chain of respiration   The oxidation–reduction reactions that use energy from the oxidation of NADH and $FADH_2$ and result in ATP synthesis from ADP and phosphate

Elongation zone (root) A region near the tip of the root where cell elongation takes place

Emerson enhancement effect The increase in photosynthetic rate that occurs when light of wavelengths shorter than 690 nm are supplemented with light of wavelengths longer than 690 nm; found to be due to the activation of the P700 reaction center of photosystem I

Endodermis The inner cells of the root cortex, which surround the stele and regulate the uptake of nutrient ions into the xylem

Endoplasmic reticulum A tubular network formed from and continuous with the nuclear envelope, which fills much of the volume of the cytosol

Enzyme A protein catalyst

Epidermis The outer layer of cells covering herbaceous plant organs like leaves, stems, and roots

Essential amino acids Those that are not made by animals and must be consumed from plant sources

Ethylene A plant hormone that is a senescence-promoting gas, also associated with fruit ripening

Etiolation Growth in the dark that results in excessive stem elongation, unexpanded leaves, and chlorosis

Exodermis (also Hypodermis) A layer of suberized cells just underneath the epidermis that retains water in mature root tissue

Fermentation The breakdown of sugar in the absence of oxygen resulting in the production of carbon dioxide (CO_2) and ethyl alcohol or lactic acid, and which uses NADH and regenerates NAD^+

Fertilization The union of an egg and a sperm to form a zygote

Fiber cells Long narrow support cells that have thick, lignified secondary cell walls; typically associated with veins

Fibrous roots Relatively thin roots produced in abundance, primarily located in upper soil layers

Filament The stalk of a stamen on which an anther is borne

Florigen The name of the protein that moves from mature leaves to the apical meristem, leading to formation of an inflorescence

Fructans Carbohydrate storage molecules consisting of one glucose molecule plus a chain of fructose molecules

Fusiform initials Meristematic cells of the vascular cambium that divide longitudinally to produce secondary xylem and secondary phloem

Gamete The egg or sperm cell in sexual reproduction, containing only one copy of genetic information

Generative cell Of the two cells in a pollen grain, the one that divides into two sperm cells. One unites with the egg to form the embryo, and the second unites with two polar nuclei (female cells in the ovule) to form the endosperm

Gibberellin A plant hormone that promotes the growth of stem internodes via cell division and elongation, and the germination of seeds; first discovered in a fungal infection of rice plants

Glycolysis The oxygen-independent reactions of respiration in the cytosol that use glucose to produce pyruvate and two molecules of ATP

Golgi apparatus A stack of flattened sacs where some cell wall polysaccharides are synthesized and glycoproteins from the endoplasmic reticulum are processed before secretion

Grana Thylakoid membranes assembled into stacks in the stroma

Gravitational water After precipitation, the water that will drain from a soil due to gravity

Gravitropism (also **Geotropism**) The growth of a plant root or shoot in response to gravity

Ground tissue The fundamental tissues that comprise the bulk of the plant

Guard cells The stomatal cells that swell or shrink to control the size of the opening of a stoma

Guttation The exudation of xylem sap from **hydathodes** at leaf tips or margins

Hardwood The wood of dicots composed of xylem elements, tracheids, parenchyma, and fiber cells

Heartwood Darkened wood in the center of a tree trunk in which tannins, resins, oils, and gums are stored; no longer able to transport water

Hemicellulose Cell wall structural carbohydrates more branched and diverse in composition than cellulose

Humus Nonliving organic matter in the soil derived from decomposing plant and animal matter

Hydathode A pore at the tip or margins of a leaf from which water and dissolved minerals may be released under xylem pressure

Hydrogen bond An attraction between two molecules due to their polarity, which influences their orientation in relation to one another

Hydrophyte A plant adapted to a wet environment; leaves typically have a thin or absent cuticle and large internal air spaces in the mesophyll (aerenchyma)

Hygroscopic water The film of water tightly adsorbed to soil mineral particles and humus and not available for uptake by plant roots

Hypodermis (see **Exodermis**)

Immobile nutrient A plant nutrient that will remain in the tissue where it was originally incorporated

Imperfect flower A flower with only staminate or pistillate flower parts, but not both

Incomplete flower A flower that is missing one of the four parts of a complete flower

Induction The response to decreasing temperatures and daylengths in fall that prepares a plant for initiation of reproductive growth in spring

Infection thread An ingrowth of a root epidermal cell plasma membrane containing *Rhizobia* that grows to the cortex ahead of a rhizobial infection

Inflorescence A group of flowers borne on the same axis or rachis

Initials Slowly dividing cells in meristems that supply new meristematic cells as needed

Initiation The change in the apical meristem from leaf production to flower production

Inner membrane The inner mitochondrial membrane with inward-directed folds (cristae) on which the reactions of the electron transport chain occur

Intercalary meristems Located at the base of leaves or internodes, such meristems support elongation of these organs

Internode Regions of the stem between nodes that elongate to increase the height of a stem

Ionic bond An attraction between ions that occurs when one atom gives up one or more electrons to another atom

Krebs cycle The series of reactions in which pyruvate is broken down to carbon dioxide (CO_2) and molecules of NADH and $FADH_2$ are produced

Lateral meristem A cylindrical meristem composed of a thin layer of meristematic cells in the stem or root of woody plants that produces radial growth (increase in girth)

Latex A milky exudate produced by some plants that contains a suspension of compounds (e.g., rubber)

Leaf primordium The earliest stage of growth of a new leaf at the apical meristem

Leghemoglobin A pigment that carries oxygen within legume root nodules

Lenticels Pores in the outer layers of bark that allow gas exchange across the cork

Lignin Simple phenolic subunits randomly bonded into large, complex molecules in the walls of fiber and xylem cells, which waterproof cells and ward off attacks by pathogens

Lipid bilayer A self-assembled membrane consisting of a double layer of phospholipids in which hydrophilic heads are turned out and hydrophobic fatty acid tails are turned inward

Lipids Fats and oils composed of glycerol and fatty acids

Lumen The space enclosed by membranes, such as the inner space of the endoplasmic reticulum

Macronutrients Those nutrients required by plants in relatively large quantities ($>0.5\%$ of dry matter)

Maturation zone A region of elongated root cells, often with root hairs, where the rate of water absorption from the soil is greatest

Megapascal (MPa) The preferred unit of water potential; 1 MPa is equal to 10 bars

Meristem A region of cell division

Mesophyll Ground tissue located between the upper and lower epidermis of leaves, which functions in photosynthesis

Micronutrients Those nutrients required by plants in relatively small quantities ($<0.5\%$ of dry matter)

Micropyle The opening into the ovule through which the pollen tube grows; the site of imbibition of water during germination

Middle lamella The outermost layer of the cell wall, containing a relatively high concentration of pectins, laid down during cell division

Mitochondrion The organelles of the cell in which the reactions of the Krebs cycle and the electron transport chain of respiration take place

Mobile nutrient A plant nutrient that can be transported from older tissue into younger tissue

Monecious plant A plant with staminate (male) and pistillate (female) flowers on the same plant (e.g., maize)

Multiple fruit A single structure formed from the fusion of several ovaries (e.g., pineapple)

Mycorrhizae Fungal organisms in mutually beneficial relationships with plant roots, with the fungal component supplying nitrogen and phosphorus to the plant and the plant supplying organic nutrients to the fungal partner

NADH and FADH$_2$ The molecules that carry reducing power produced primarily by the Krebs cycle of respiration

Nectary Special glands at the base of the petals, pistil, or stamens that secrete nectar in order to attract pollinators to a flower

Nitrogenase The nitrogen-fixing enzyme in rhizobia bacteria, active in legume nodules

Node The location on the stem where leaves and axillary buds are attached

N–P–K ratio The ratio of nitrogen, phosphorus (as P_2O_5), and potassium (as K_2O) in a fertilizer formulation

Nucleic acid Linear polymers composed of nucleotides. In nucleic acids, phosphate alternates with either ribose (RNA) or deoxyribose (DNA) to which a base is attached. Nucleic acids store or translate genetic information into proteins.

Nucleotides The subunits of DNA and RNA, composed of a sugar, a base, and a phosphate

Nucleus Found in most living cells, the location of genetic material (DNA) which serves as a template for the synthesis of cellular proteins via RNA

Organic acid Carbohydrates such as citrate and malate that readily lose a proton (H^+) and thereby gain the ability to combine with cations

Organic molecules Molecules composed of carbon, hydrogen, and usually oxygen that are involved in the biochemistry of living organisms

Organogenesis The formation of roots or shoots in nature or cultured plant tissue

Osmosis Equalization of the concentration of water molecules requiring diffusion of water across a differentially permeable membrane

Ovary The basal portion of a pistil that develops into a fruit, and which may contain more than one ovule or seed

Oxidative phosphorylation The phosphorylation of ADP to ATP that occurs in respiration as a result of the series of oxidation and reduction reactions in the electron transport chain. The carbohydrate starting material (glucose) is completely oxidized to water in this process

P680 The reaction center of photosystem II that absorbs light at 680 nm

P700 The reaction center of photosystem I that absorbs light at 700 nm

Palisade parenchyma (mesophyll) cells Closely packed, elongated leaf mesophyll cells located directly below the upper epidermis of dicot leaves

Palmate leaf A compound leaf in which leaflets are attached to the petiole at a single point

Panicle An inflorescence whose flowers are borne on pedicels attached to branches that are in turn attached to the rachis

Parallel venation The venation of a monocot leaf in which veins are continuous and parallel from base to tip

Parenchyma Ground tissue that functions in storage or metabolism, including the cortex and pith

Peduncle The extension of a stem or branch that ends in a flower or inflorescence

Perfect flower A flower that has both staminate (male) and pistillate (female) flower parts

Perforation plates The large openings in the adjacent end walls of vessel elements

Perianth A collective term for the outer structures of a flower, usually consisting of the calyx and corolla

Pericycle A layer of cells located just inside the endodermis of the root that is the site of branch root formation

Permanent wilting point A soil water potential so low that plants cannot take up water; about -1.5 MPa

Petal Usually brightly colored flower parts surrounding the pistil and stamens that assist in attracting pollinators

Petiole A stalk that attaches a leaf to a stem, and which is absent in a sessile leaf

pH A measure of the acidity or alkalinity of a solution

Phenolics A group of secondary compounds containing aromatic rings; derived from carbohydrates by the shikimic acid pathway

Phloem Vascular tissue responsible for the transport of sugars and other organic and inorganic nutrients

Phosphoenolpyruvate carboxylase (PEPcase) The enzyme that adds a molecule of CO_2 to phosphoenol pyruvate in mesophyll cells, creating a 4-carbon acid

3-Phosphoglyceric acid (3-PGA) The 3-carbon sugar that is the first stable product of the Calvin cycle of photosynthesis

Phospholipid The dominant component of the lipid bilayer of the plasma membrane, with a hydrophilic (water-loving) phosphate-containing head and a hydrophobic (water-fearing) tail composed of two fatty acids

Photochemical (light) reactions of photosynthesis The reactions in which the energy of the sun is captured by chlorophyll and used to create NADPH and ATP

Photoperiodism The development of an inflorescence in response to change in the relative lengths of day and night

Photophosphorylation The synthesis of ATP that occurs as a by-product of the light reactions of photosynthesis

Photorespiration The loss of a CO_2 due to the addition of two oxygen molecules (O_2) to RuBP by RuBP carboxylase/oxygenase (rubisco)

Photosynthate The carbohydrates made via photosynthesis

Photosynthesis The process by which plants and some other organisms capture the energy of light and convert it into chemical energy in the form of sugar

Phototropism Curvature of a dark-grown plant shoot toward light

Phytochrome The pigment that serves as the biological switch in many light-sensitive physiological responses of plants

Pinnate leaf A compound leaf with leaflets attached along the central axis of the leaf

Pistil The female sexual column of a flower composed of at least one stigma, style, and ovary

Pith A tissue located in the center of the stem or root composed of thin-walled cells

Plasma membrane The outer membrane of a plant cell, composed of a lipid bilayer in which metabolically active protein complexes, such as those that control the movement of dissolved substances into and out of the cell, are imbedded

Plasmalemma (see **Plasma membrane**)

Plasmodesmata Plasma membrane-lined channels between cells through which cytosol can pass but which are too small for cellular organelles

Polar molecule A molecule that carries equal partial positive and partial negative charges

Polar movement Movement in a specific direction

Pollen Fine, dust-like structures of a unique form, developed in anthers, and containing sperm (male gametes)

Pollen tube A channel that grows from the stigma through the style and into the ovule, carrying the generative cell from the pollen grain which divides into two sperm cells during migration

Pollination The transfer of a pollen grain from an anther to a receptive stigma of the same species

Primary cell wall A rigid cellular structure composed largely of cellulose microfibrils that supports and restricts the cellular contents

Primary growth Growth that results in an increase in length or height of the root or shoot

Primary protein structure The sequence of amino acids in a protein

Proteins Polymers of amino acids that are linked through peptide bonds

Protoplast The plasma membrane and its living cellular contents, including the nucleus

Pubescence Hairiness that may protect against insects, reflect heat, or decrease air movement at the leaf surface

Quaternary protein structure An assembly of protein subunits constituting a single functioning protein

Quiescent center A region of meristem initials in root tips

Raceme An inflorescence whose flowers are attached to the rachis by pedicels

Rachis The central axis of an inflorescence

Reaction center A chlorophyll a (chl a) molecule that loses an electron to the electron transport chain, activated by energy from the antenna complex

Receptacle A swollen region at the base of a flower to which other flower parts are attached

Resin A sticky aromatic fluid that oozes from specialized canals when the canals are broken, and hardens on exposure to air

Respiration The sequence of steps whereby energy stored primarily in carbohydrates is used to synthesize ATP that can be used to drive chemical reactions and other processes

Reticulate venation The net-like venation of a dicot leaf

Rhizobia Soil bacteria that can become established in legume roots where they convert nitrogen gas (N_2) from the atmosphere to a form useful to plants (NH_4^+)

Rhizomes Belowground horizontal stems that originate from axillary buds, which can produce new plants vegetatively

Ribosomes Cellular structures composed of RNA and protein that are the site of protein synthesis; found free in the cytosol or attached to the endoplasmic reticulum

Ribulose bisphosphate carboxylase/oxygenase (Rubisco) The enzyme that catalyzes the first step of the dark reactions of photosynthesis, adding CO_2 to ribulose bisphosphate (RuBP); rubisco can also add O_2 to RuBP in C_3 plants, resulting in photorespiration

Root cap Cells that cover and protect the root tip, producing mucigel to lubricate the soil

Root nodules Tumorous growths on plant roots invaded by *Rhizobia* bacteria in which nitrogen fixation takes place

Root pressure Pressure in the xylem sap resulting from the uptake of mineral ions or the breakdown of starch in excess of that needed to maintain turgidity

Rosette A stem in which internodes are unelongated and leaves are arranged at ground level with minimal overlap

Sand Relatively large soil mineral particles, ranging from 0.02 to 2 mm in diameter

Sapwood The water-carrying wood of a tree trunk

Sclerenchyma Support tissue that has thickened, lignified secondary cell walls

Sclerid A sclerenchyma cell that is short and variable in shape

Sclerophyllous leaves Highly fibrous "evergreen" leaves that are sufficiently durable to be retained by the plant for more than one season

Secondary cell wall The thick, lignified inner wall layers that form in fiber cells to provide the skeletal structure of the plant, or in xylem transport cells to reinforce the cell against the pull of transpiration

Secondary growth Growth that results in increase in the girth of woody stems or roots

Secondary phloem The tissue that transports sugars along the outside of the stem in woody plants, composed of fibers, sieve elements, companion cells, and parenchyma storage cells

Secondary plant products The products of biochemical pathways that are not essential to the basic metabolism of plants

Secondary protein structure The twisting and folding of linear sequences of proteins that results from interactions between hydrogen and oxygen and hydrogen and nitrogen atoms (hydrogen bonds)

Secondary xylem The cells that transport water and mineral nutrients through the stem in woody plants

Senescence The removal of proteins, ions, and other soluble material from a plant organ as a normal function of aging, followed by death

Sepals Bracts subtending the petals that enclose a developing flower

Sessile leaf A leaf with no petiole, attached directly to the stem

Shade leaf A leaf adapted to capture sunlight for photosynthesis in a shady environment; thinner and with more surface area than a sun leaf of the same plant

Sieve area The end walls of sieve elements perforated by sieve pores for rapid cell-to-cell movement of nutrients

Sieve elements Cells of the phloem with no nucleus or vacuole, through which sugars and other nutrients are carried; living at maturity

Silt Intermediately sized soil mineral particles, ranging from 0.02 to 0.002 mm

Simple leaf A leaf with an undivided blade

Sink An organ or tissue that requires photosynthate but either does not produce any or does not produce enough to meet its own needs

Softwood The lightweight wood of conifers consisting of tracheids but no vessel elements

Solute Substances that become dissolved

Solvent A liquid capable of dissolving or dispersing other substances

Source A photosynthetic tissue that produces more photosynthate than it requires for its own use

Sperm cells The male gametes (reproductive cells); one unites with the egg and the other unites with two polar nuclei in the ovule

Spike An inflorescence whose flowers have no pedicels, but attach directly to the rachis

Spines Modified leaves such as the rigid remnants of petioles and midribs that occur on the stems of cacti

Spongy mesophyll cells Loosely packed mesophyll cells in the lower part of dicot leaves with extensive air spaces for gas exchange (aerenchyma)

Springwood Large-diameter secondary xylem formed during leaf expansion in spring, when water is usually plentiful

Stamen An anther and its filament

Starch Carbohydrate storage molecules consisting of long chains of glucose molecules

Stele The inner region of the root, inside the endodermis, containing the vascular tissue

Stigma A sticky surface at the top of the pistil to which pollen adheres

Stipules Leaf-like appendages at the junction of the stem and leaf, which enclose developing leaves and axillary buds

Stolons Aboveground horizontal stems that originate from axillary buds, which can produce new plants vegetatively

Stomata Microscopic pores in the surface of the leaf that provide the gas exchange needed for photosynthesis

Stratification Seed imbibition and cold temperature storage that promotes germination

Stroma The semiliquid matrix of chloroplasts in which the dark reactions of photosynthesis occur

Style The stalk of the ovary through which a pollen tube must grow for fertilization to occur within each ovule

Suberin A waxy substance that waterproofs cork cells and the Casparian strip of the endodermis

Subsidiary cells Stomatal cells that support the function of guard cells

Summerwood Small-diameter secondary xylem formed when leaves are fully expanded and water is usually restricted

Sun leaf A dicot leaf adapted to full sun, with multilayered palisade parenchyma dense with small chloroplasts; smaller and thicker than shade leaves of the same plant

Symbiotic Mutually beneficial

Symplast The continuous cytoplasmic (living) network among cells through which mineral nutrients and other dissolved substances can travel from cell to cell

Tannins A group of phenolic secondary compounds characterized by their ability to bind proteins; long used to inhibit microbial degradation of animals skins

Taproot A main root that grows downward to explore the soil, sometimes to great depths; the taproots of some perennial plants store organic nutrients

Tendrils Modified leaves or leaflets that grow with a circular motion to contact external support

Tensile strength The tension (pull) a continuous column of a material can withstand before it breaks

Terpenes A group of lipid-like secondary compounds built from 5-carbon isoprene subunits

Tertiary protein structure The three-dimensional structure dictated by interactions among amino acid side groups and stabilized by occasional covalent disulfide bonds

Thigmomorphogenesis The stunting of plants exposed to mechanical disturbance due to the hormone ethylene

Thigmotropism The response of plants to touch that results in the ability of tendrils to grasp objects

Thylakoid membranes The internal membranes of chloroplasts, organized into flattened vesicles, in which chlorophyll and the carotenoids are embedded

Tillers (also **Ramets**) Shoots arising from the axillary buds of grasses; a form of vegetative or clonal reproduction

Tonoplast The membrane enclosing the vacuole

Tracheids Nonliving water-carrying cells of the xylem, especially in gymnosperms, with wall thickenings, narrow diameter, and tapered ends that rely on pit pairs for water movement

Translocation The movement of substances from one location in the plant to another, used especially in reference to movement of carbohydrates through the phloem

Transpiration The loss of water vapor from the saturated internal air spaces of leaves through open stomata to the air around leaves

Trapping leaves Leaves that can trap and digest insects to harvest nitrogen

Trichomes Hairs, glands, and other structures of varying complexity that arise from the epidermis

Tropism Growth in response to a stimulus, such as light or gravity

Tube cell Of the two cells in a pollen grain, the one that forms the pollen tube

Tuber Stem tissue modified for storage that can be used for vegetative propagation if it contains an axillary bud

Turgid Fully inflated, as when the vacuole of a cell expands and presses the cytoplasm against the cell wall

Turgor pressure The positive pressure that develops as water is drawn into a cell by a decrease in osmotic potential

Twining shoots Modified stems that grow with a circular motion to contact external support

Umbel An inflorescence in which the pedicels are attached to the peduncle near one point

Vacuole The organelle that occupies much of the space in plant cells, generally filled with water and used for storage of mineral nutrients and other dissolved substances

Vascular cambium A cylindrical meristem located between the secondary xylem and secondary phloem in woody stems that is the source of new xylem and phloem tissue and which results in secondary growth (increase in girth)

Vegetative clone A plant that is identical to the mother plant because it has developed from an axillary bud or root sucker rather than through sexual reproduction

Vernalization Exposure to temperatures close to freezing that promotes flowering

Vessel elements Nonliving water-carrying cells of the xylem of angiosperms with relatively large diameter that have both pit pairs and perforated end plates for water movement

Vivipary The germination of immature seeds on the seedhead due to a deficiency of abscisic acid

Water-holding capacity The water content of a thoroughly wetted soil after surplus water has drained by gravity

Water potential The force that causes water to flow from locations where it is relatively more pure or where it is under pressure to locations where water is under less pressure or where it has a higher solute concentration

Xerophyte A plant adapted to dry climates; leaves are typically thick and small in diameter, with multilayered upper epidermis, thick cuticles, and sunken stomata

Xylem Vascular tissue responsible for the transport of water and mineral nutrients

Zeatin The most common plant cytokinin; originally purified from maize (*Zea mays*)

Zygote The single cell containing one complete set of chromosome pairs, half from the male parent and half from the female parent, which will develop into the embryo

References

Bennett, W. F. 1993. *Nutrient Deficiencies & Toxicities in Crop Plants*. American Phytopathological Society, St. Paul, MN.

Biale, J. B. 1964. Growth, maturation, and senescence in fruits. Recent knowledge on growth regulation and on biological oxidations has been applied to studies with fruits. *Science* 146:880–888.

Bjørkman, O., H. A. Mooney, and J. Ehleringer. 1975. *Photosynthetic responses of plants from habitats with contrasting thermal environments: comparison of photosynthetic characteristics of intact plants*. Yearbook 74 (1974–1975), Carnegie Institution of Washington, Washington, DC, pp. 743–748.

Blevins, D. G. 1984. The potassium–protein link. Why some crops need more K than others. *Crops and Soils Magazine* 36:12–13.

Borthwick, H. A., S. B. Hendricks, and M. W. Parker. 1952a. The reaction controlling floral initiation. *Proceedings of the National Academy of Sciences of the United States of America* 38:929–934.

Borthwick, H. A., S. B. Hendricks, M. W. Parker, E. H. Toole, and V. K. Toole. 1952b. A reversible photoreaction controlling seed germination. *Proceedings of the National Academy of Sciences of the United States of America* 38:662–666.

Boyd, R. S. 1994. Pollination biology of the rare shrub *Fremontodendron decumbens* (Sterculiaceae). *Madroño* 41:277–289.

Boyd, R. S. 1996. Ant-mediated seed dispersal of the rare chaparral shrub *Fremontodendron decumbens* (Sterculiaceae). *Madroño* 43:299–315.

Boyd, R. S. 2001. Ecological benefits of myrmecochory for the endangered chaparral shrub *Fremontodendron decumbens* (Sterculiaceae). *American Journal of Botany* 88:234–241.

Boyd, R. S. and L. L. Serafini. 1992. Reproductive attrition in the rare chaparral shrub *Fremontodendron decumbens* Lloyd (Sterculiaceae). *American Journal of Botany* 79:1264–1272.

Boyer, P. D. 1993. The binding change mechanism for ATP synthase—some probabilities and possibilities. *Biochimica et Biophysica Acta* 1140:215–250.

Brady, N. 1990. *The Nature and Property of Soils*, 10th Ed. Pearson, Upper Saddle River, NJ.

Brown, H. P., A. J. Panshin, and C. C. Forsaith. 1949. *Textbook of Wood Technology*, Vol. 1. McGraw-Hill, New York.

Butler, W. L. 1964. Introduction. Symposium on photomorphogenesis in plants. *The Quarterly Review of Biology* 39:1–5.

Butler, W. L., K. H. Norris, H. W. Siegelman, and S. G. Hendricks. 1959. Detection, assay, and preliminary purification of the pigment controlling photoresponsive development of plants. *Proceedings of the National Academy of Sciences of the United States of America* 45:1703–1708.

Caldwell, M. M. 1988. Plant root systems and competition. In W. Greuter and B. Zimmer (eds), *Proceedings of the XIV International Botanical Congress*, Berlin,

July 24 to August 1, 1987. Koeltz Scientific Books, Königstein, Germany, pp. 385–404.

Calvin, M. 1959. Energy reception and transfer in photosynthesis. *Reviews of Modern Physics* 31:147–156.

Chailakhian, M. K. 1961. Principles of ontogenesis and physiology of flowering in higher plants. *Canadian Journal of Botany* 39:1817–1841.

Chouard, P. 1960. Vernalization and its relations to dormancy. *Annual Review of Plant Physiology* 11:191–238.

Corbesier, L., C. Vincent, S. Jang, F. Fornara, Q. Fan, I. Searle, A. Giakountis, S. Farrona, L. Gissot, C. Turnbull, and G. Coupland. 2007. FT protein movement contributes to long-distance signaling in floral induction of Arabidopsis. *Science* 316:1030–1033.

Dhont, C., Y. Castonguay, P. Nadeau, G. Bélanger, and F.-P. Chilifour. 2002. Alfalfa root carbohydrates and regrowth potential in response to fall harvests. *Crop Science* 42:754–765.

Dhont, C., Y. Castonguay, P. Nadeau, G. Bélanger, and F.-P. Chilifour. 2003. Alfalfa root nitrogen reserves and regrowth potential in response to fall harvests. *Crop Science* 43:181–194.

Dittmer, H. J. 1937. A quantitative study of the roots and root hairs of a winter rye plant (*Secale cereale*). *American Journal of Botany* 24:417–420.

Dittmer, H. J. 1938. A quantitative study of the subterranean members of three field grasses. *American Journal of Botany* 25:654–657.

Dodds, D. 1980. *The Growth of Grasses*. Cooperative Extension Service, North Dakota State University, Fargo, ND.

Donnelly, P. M., D. Bonetta, H. Tsukaya, R. E. Dengler, and N. G. Dengler. 1999. Cell cycling and cell enlargement in developing leaves of *Arabidopsis*. *Developmental Biology* 215:407–419.

Drew, M. C., M. B. Jackson, and S. Giffard. 1979. Ethylene-promoted adventitious rooting and development of cortical air spaces (aerenchyma) in roots may be adaptive responses to flooding in *Zea mays* L. *Planta* 147:83–88.

Dugger, B. M. 1924. *Plant Physiology with Special Reference to Plant Production*. Macmillan, New York.

Epstein, E., and A. J. Bloom. 2005. *Mineral Nutrition of Plants: Principles and Perspectives*, 2nd Ed. Sinauer, Sunderland, MA.

Feild, T. S., D. W. Lee, and N. M. Holbrook. 2001. Why leaves turn red in autumn. The role of anthocyanins in senescing leaves of red-osier dogwood. *Plant Physiology* 127:566–574.

Garner, W. W., and H. A. Allard. 1920. Effect of the relative length of day and night and other factors of the environment on growth and reproduction in plants. *Journal of Agricultural Research* 18:553–606.

Graham, J., D. T. Clarkson, and J. Sanderson. 1974. *Water Uptake by the Roots of Marrow and Barley Plants*. Annual Report 1973, Letcombe Laboratory, Wantage. Agricultural Research Council, London.

Hamilton, A. J., G. W. Lycett, and D. Grierson. 1990. Antisense gene that inhibits synthesis of the hormone ethylene in transgenic plants. *Nature* 346:284–287.

Holmes, M. G. and H. Smith. 1975. The function of phytochrome in plants growing in the natural environment. *Nature* 254:512–514.

Hopkins, W. G. 1995. *Introduction to Plant Physiology*. Wiley, New York.

James, D. W., R. J. Hanks, and J. J. Jurinak. 1982. *Modern Irrigated Soils*. Wiley, New York.

Jung, G. A. and D. Smith. 1961. Trends of cold resistance and chemical changes over winter in the roots and crowns of alfalfa and medium red clover. I. Changes in certain nitrogen and carbohydrate fractions. *Agronomy Journal* 53:359–366.

Juniper, B. E., S. Groves, B. Landau-Schachar, and L. J. Audus. 1966. Root cap and the perception of gravity. *Nature* 209:93–94.

Kimball, J. W. 1983. *Biology*, 5th Ed. Addison-Wesley, Reading, MA.

Lauer, M. J., D. G. Blevins, and H. Sierzputowska-Gracz. 1989. ^{31}P-Nuclear magnetic resonance determination of phosphate compartmentation in leaves of reproductive soybeans (*Glycine max* L.) as affected by phosphate nutrition. *Plant Physiology* 89:1331–1336.

Ledbetter, M. C., and K. R. Porter. 1970. *Introduction to the Fine Structure of Plant Cells*. Springer-Verlag, New York.

Leopold, A. C. and M. Kawase. 1964. Benzyladenine effects on bean leaf growth and senescence. *American Journal of Botany* 51:294–298.

Luckwill, L. C. 1952. Growth-inhibiting and growth-promoting substances in relation to the dormancy and after-ripening of apple seeds. *Journal of Horticultural Science* 27:53–67.

Martin, J. H., W. H. Leonard, and D. L. Stamp. 1976. *Principles of Field Crop Production*, 3rd Ed. Macmillan, New York.

McGuire, V. L. 2004. Water-level changes in the High Plains aquifer, predevelopment to 2003 and 2002 to 2003. USGS Fact Sheet 2004-3097. Available at http://pubs.usgs.gov/fs/2004/3097/.

Moll, S., S. Anke, U. Kahmann, R. Hänsch, T. Hartmann, and D. Ober. 2002. Cell-specific expression of homospermidine synthase, the entry enzyme of the pyrrolizidine alkaloid pathway in *Senecio vernalis*, in comparison with its ancestor, deoxyhypusine synthase. *Plant Physiology* 130:47–57.

Nativ, R. and D. A. Smith. 1987. Hydrogeology and geochemistry is the Ogallala aquifer, southern High Plains. *Journal of Hydrology* 91:217–253.

Neales, T. F., A. A. Patterson, and V. J. Hartney. 1968. Physiological adaptation to drought in the carbon assimilation and water loss of xerophytes. *Nature* 219:469–472.

Nelson, C. J. and L. E. Moser. 1995. Morphology and systematics, in *Forages: An Introduction to Grassland Agriculture*, 5th Ed., Vol. I, pp. 15–30, R. F. Barnes, D. A. Miller, and C. J. Nelson (eds). Iowa State University Press, Ames, IA.

Noji, H., R. Yasuda, M. Yoshida, and K. Kinosita. 1997. Direct observation of the rotation of F_1-ATPase. *Nature* 386:299–302.

Noji, H. and M. Yoshida. 2001. The rotary machine in the cell, ATP synthase. *The Journal of Biological Chemistry* 276:1665–1668.

Pearce, R. S. 1988. Extracellular ice and cell shape in frost-stressed cereal leaves: a low-temperature scanning-electron-microscopy study. *Planta* 175:313–324.

Pfister, J. A., K. E. Panter, D. R. Gardner, B. L. Stegelmeier, M. H. Ralphs, R. J. Molyneux, and S. T. Lee. 2001. Alkaloids as anti-quality factors in plants on western U.S. rangelands. *Journal of Range Management* 54:447–461.

Pimentel, D., J. Houser, E. Preiss, O. White, H. Fang, L. Mesnick, T. Barsky, S. Tariche, J. Schreck, and S. Alpert. 1997. Water resources: agriculture, the environment, and society. An assessment of the status of water resources. *BioScience* 47:97–106.

Pool, R. J. 1923. Xerophytism and comparative leaf anatomy in relation to transpiring power. *Botanical Gazette* 76:221–240.

Rastorfer, J. R. and N. Higinbotham. 1968. Rates of photosynthesis and respiration of the moss *Bryum sandbergii* as influenced by light intensity and temperature. *American Journal of Botany* 55:1225–1229.

Rosenberg, N. J., D. J. Epstein, D. Wang, L. Vail, R. Srinivasan, and J. G. Arnold. 1999. Possible impacts of global warming on the hydrology of the Ogallala aquifer region. *Climatic Change* 42:677–692.

Schaberg, P. G., A. K. Van Den Berg, P. F. Murakami, J. B. Shane, and J. R. Donnelly. 2003. Factors influencing red expression in autumn foliage of sugar maple trees. *Tree Physiology* 23:325–333.

Shaw, J. 2004. *Trans Fats The Hidden Killer in Our Food*. Pocket Books, New York.

Slatyer, R. O. 1967. *Plant Water Relationships*. Academic Press (Elsevier), New York.

Smith, C. J. S., C. F. Watson, J. Ray, C. R. Bird, P. C. Morris, W. Schuch, and D. Grierson. 1988. Antisense RNA inhibition of polygalacturonase gene expression in transgenic tomatoes. *Nature* 334:724–726.

Smith, D., R. J. Bula, and R. P. Walgenbach. 1986. *Forage Management*, 5th Ed. Kendall Hunt, Dubuque, IA.

Smith, H. and D. C. Morgan. 1981. The spectral characteristics of the visible radiation incident upon the surface of the Earth, in *Plants and the Daylight Spectrum*, pp. 3–20, H. Smith (ed). Academic Press, London.

Stryer, L. 1975. *Biochemistry*, 2nd Ed. Freeman, New York.

Taiz, L. and E. Zeiger. 2006. *Plant Physiology*, 4th Ed. Sinauer, Sunderland, MA.

Talbott, L. D., and E. Zeiger. 1998. The role of sucrose in guard cell osmoregulation. *Journal of Experimental Botany* 49:329–337.

Tamaki, S., S. Matsuo, H. L. Wong, S. Yokoi, and K. Shimamoto. 2007. Hd3 a protein is a mobile flowering signal in rice. *Science* 316:1033–1036.

Thorup, R. M. 1984. *Ortho Agronomy Handbook. A Practical Guide to Soil fertility and Fertilizer Use*. Chevron Chemical Company, San Francisco, CA.

Trapp, S. and R. Croteau. 2001. Defensive resin biosynthesis in conifers. *Annual Review of Plant Physiology and Plant Molecular Biology* 52:689–724.

Trewartha, G. T., and L. H. Horn. 1980. *An Introduction to Climate*, 5th Ed. McGraw-Hill, New York.

Turgeon, R. and J. A. Webb. 1973. Leaf development and phloem transport in *Cucurbita pepo*: transition from import to export. *Planta* 113:179–191.

USDA, NRCS. 2007. The PLANTS Database, available at http://plants.usda.gov, December 27, 2007. National Plant Data Center, Baton Rouge, LA.

Volenec, J. J. and C. J. Nelson. 2003. Environmental aspects of forage management, in *Forages: An Introduction to Grassland Agriculture*, 6th Ed., Vol. I, pp. 99–124, R. F. Barnes, C. J. Nelson, M. Collins, and K. J. Moore (eds). Iowa State Press, Ames, IA.

Wardlaw, I. F. 1969. The effect of water stress on translocation in relation to photosynthesis and growth, II. Effect during leaf development in *Lolium temulentum* L. *Australian Journal of Biological Sciences* 22:1–16.

Weaver, J. E. 1926. *Root Development of Field Crops*. McGraw-Hill, New York.

Weaver, J. E. and W. E. Bruner. 1927. *Root Development of Vegetable Crops*. McGraw-Hill, New York.

Wilkins, M. 1988. *Plantwatching: How Plants Remember, Tell Time, Form Relations and More*. Facts on File, New York.

Zimmermann, M. H. 1963. How sap moves in trees. *Scientific American* 208:132–142.

Index

ABA–see Abscisic acid
Abiotic influences 1
Abscisic acid (ABA) 135, **136**, 215, 230, 232, 233, 241
Abscisin 233
Abscission 114, 221, **230**
Absorption spectrum 160, 163
ACC (1-aminocyclopropane-1-carboxylic acid) 229
Accessory pigment **163**
Acetic acid 186
Acetyl Co-A (Acetyl coenzyme A) 186
Achene **102**
Acidity **110**
Actin filament **192**, 193
Action spectrum of photosynthesis **160**, 162, 163
Action spectrum of phytochrome **201**, 204
Activation energy 154
Activator, enzyme 156
Active site, enzyme 154, 155, 156, **157**, 171, 191
Active transport 126, 129, 216
Active uptake **126**, 127
Adams, Thomas 240
Adenine 147, **148**, 149, **183**, 226
Adenosine diphosphate–see ADP
Adenosine monophosphate–see AMP
Adenosine triphosphate–see ATP
ADP (adenosine diphosphate) 165, 166, 182, **183**, **190**, **191**
Adrenaline 248
Adventitious root 37, **38**, 219, **232**
Aeration, soil 106, 107
Aerenchyma **85**, 231, **232**
Aerobic respiration 186
Aggregate fruit 91
Aleurone layer **225**, 226
Alfalfa **99**, **133**, 243
Alkalinity **110**
Alkaloids 235, 248, **249**, 251
Allard, Henry 195

Allelopathy 242
Amino acid 149, **150**, **151**
Amino group 149, **150**, **151**, **153**
Ammonia (NH_3) 49, **53**, 122
Ammonium (NH_4^+) 116
AMP (Adenosine monophosphate) 176
α-Amylase 225, **226**, 234
Amyloplast 15, 21, **22**, 218, **219**, **220**
Amylose 225
Analgesic 248
Angiosperm 30, 89
Anion 111, **142**
Annual rings 65
Antenna complex **163**, **164**, 165
Anther 91, **101**, 221
Anthocyanins 26, 229, **244**, **245**, **246**
Apical dominance 219, **221**, 226
Apical meristem 18, **19**, **20**, **62**, 89, 202, 207, 218, 221, 226
Apoplast 11, **47**, 113, 127, 129, **212**
Apoplastic movement 87, **113**
Aquaporin **126**
Aquifer 132
Aspartate **175**
Astringency 247
ATP (adenosine triphosphate) 12, 13, 15, 53, 88, 118, 126, 137, 147, **156**, 158, **159**, 165, 166, **167**, 170, 172, 175, 176, 177, 180, 182, **183**, 184, 185, 186, **187**, 188, 190, 191, **192**, 193
ATP synthase 166, **167**, 190, **191**, 192
Autoradiography **86**, 170, 171
Autotoxicity **243**
Auxin 215, 216, 218, **220**, 222, 230, 231
Axillary bud **19**, 20, 40, 55, 57, 89, 202, 218, 226, 227

Bacteroids **51**, 52, 53
Bark 66, 67, **69**
Bases, nitrogenous 147, **148**, 149
Bee **98**
Benzene ring **242**

Benzyladenine **228**
Berry **102**
Biennial plants 36, **223**
Biloxi soybeans 195, **196**
Biochemical pathway 155
Biochemical reactions **155**, **156**
Biochemical reactions of photosynthesis 136, **158**, **159**, **165**, **166**, **168**, **170**, **171**, **177**
Biochemistry 141
Biotic influences 1
Biotin 192
Bipedal 176
1,3-Bisphosphoglycerate (BPG) **172**
Blade, leaf 19, 72, **73**, **74**
Bloat 248
Blossom-end rot 117
Bolting **223**
Bonsai 34
Bordered pit 30, **31**, **32**
Borlaug, Norman 225
Boron (B) 121
Brace root 38
Bract 94
Branch root 38, 47
Buffalograss 93
Bulb 38, 59, **60**
Bulb scale **22**, 77
Bulliform cell 27, **29**, 127
Bunchgrass 58
Bundle sheath **174**, **175**, **180**
Burning bush **246**

C_3 photosynthesis 167, **175**, **177**
C_3 species **134**, **174**, **176**
C_4 photosynthesis 121, **174**, **175**, **176**
C_4 species **174**, **175**, **176**
Cacao 249
Cacti, cactus 55, **57**, 77
Caffeine 248, **249**
Calcium (Ca) 12, 117, 165
Calcium carbonate 110
Callus 227
Calvin cycle 171, **172**, **174**, **175**, **177**, **178**
Calvin, Melvin 168, 171
Calyx 89
CAM photosynthesis 121, **134**
CAM species **177**
Canopy 133, 204, **207**
Capillary water **109**
Carbohydrate 13, 134
Carbon dioxide (CO_2) 10, 71

Carboxyl group 146, 149, **150**, **151**
Carboxylase 171
Carboxylation 172
Carboxylic acid 186, **187**
β-Carotene **162**, 241
Carotenoid 160, **163**, 229, 241, 244
Casparian strip **47**
Catalyst, enzyme 155
Cation 111, **112**, 142
Cation exchange capacity–see CEC
CEC 112
Celery 224
Cell division 6, 8, 81, **179**, 222, 226, 234
Cell division zone 41, **42**
Cell elongation **42**, 216
Cell growth **127**
Cell junction 23
Cell plate 8
Cell wall 2, 8, 121
Cell wall growth **127**
Cellulase 229, 230
Cellulose 3, **4**, **5**, 7
Cellulose synthase **6**, 11
Central mother cells 20
CF_0 **167**
CF_1 **167**
Channel, protein **126**
Charge deficit **126**
Chelate 120
Chicle 240
Chilling stress 209
Chloride (Cl⁻) 13, 233
Chlorine (Cl) 119, 165
Chlorophyll 15, 114, 117, 120, 229, 244
Chlorophyll a (chl a) **160**, **161**, **162**, **163**
Chlorophyll b (chl b) **160**, **162**, **163**
Chloroplast **15**, **16**, **29**, 88, **159**, **177**, **181**, **182**, 209
Chlorosis 114, 117, 118
Chocolate 248
Chromatography, paper **169**
Chromoplast 15
Chromosomes 6, 12
Chrysanthemum 224, **225**
Cis double bond **146**
Cisternae 13
Citric acid 186, **187**, 188
Citric acid cycle 186, **187**
Civil War 251
Clay 106, **107**
Climacteric **179**, 229, 230

CO_2 (carbon dioxide) 124, 133, **134**, 135, **136**, **137**, 158, 171, **173**, **174**, 175, 176, **177**, 178, 186, **187**, 188, **193**
$^{14}CO_2$ **168**
Coca 251
Cocaine **249**, 251
Codeine 249, 251
Coenzyme A 118, 186
Coffee 248
Cohesion 138, **139**
Coleoptile 204, 217
Collenchyma **23**
Colorado Plateau 251
Columbine **95**
Companion cell 32, **33**
Compensatory growth 136, **137**
Competitive inhibitor 156
Complementary strand 148
Complete flower 89
Composite head flower 95, **96**
Compound leaf 71, **73**, 79
Compound umbel **96**
Congestive heart failure 241
Conifer 38, **80**
Contractile root 38, **40**
Copper (Cu) 121
Cork 68, **69**, 70
Cork cambium 21, 62, 68
Cork oak **70**
Corm 59, **60**
Corolla 89
Corolla tube 100
Cortex 21, 46, **47**
Cotyledon 225, **226**
Cotyledonary node **40**
Covalent bond 4, 141, **142**, 147, 154
Crassulacean acid metabolism–see CAM
Cristae **13**
Critical photoperiod 200
Crop coefficient **133**
Cross-pollination 98
Cuticle 25, **26**
Cutin 25
Cyclopamine 253
Cysteine 118, 152
Cytochrome 154
Cytokinin 215, 226, **227**, 228
Cytoplasm 1, 12, **25**
Cytoplasmic membrane 8
Cytosine **148**, **149**, 226
Cytosol 1, 12, 152, **153**, 180, 182

2,4-D 220, **222**
Dahlia 60
Dandelion 220, 222
Dark reactions of photosynthesis–see Biochemical reactions of photosynthesis
Darwin, Charles 217
Daylength 195, 198, **199**
Day-neutral plant 196
Defoliation 229
Dehydration reaction **143**, **144**, **145**, 146, 150, 183
Deoxyribonucleic acid–see DNA
Deoxyribose 147, **148**, 149
Desmotubule **11**
Diarrhea 251
Dicot 7, 38, 78, **81**, 83
Dictyosome–see Golgi apparatus
Differentially permeable membrane 124, **125**
Diffuse-porous wood **66**
Diffusion 124, **125**
Digitalin 241
Diglyceride 10
Dioecious species 92, **93**
Disk flower 95, **96**
Dispersal mechanism 100, **102**
Disulfide bond or bridge 152, **154**
Diterpene 236
DNA (deoxyribonucleic acid) 12, 147, **148**, **149**, **153**
Dormancy 208, 209, **211**, 234
Dormin 233
Double helix 148, **149**
Dropsy 241
Drought 30, 65, 84, 87, **135**, **136**, **137**, 171, 178, 208, 220, 230, 233, 238, 239
Dwarf plants **223**

Efflux carrier 216
Egg 91, 100, **101**
Elaiosome 103, **104**
Electrolyte **214**
Electron **137**, 141, **142**, 164, **165**, **167**, 189, 194
Electron transport chain 11, 118, 120, 121, 154, **164**, **165**, **167**, 180, 184, 186, 188, **189**, **190**, **191**, **193**, 194
Elongation zone **42**, 218, **220**
Embryo 91, 100, **101**, 234
Emerson enhancement effect 163, 165
Endodermis 43, **47**, 78, 112, **113**, 129

Endoplasmic reticulum (ER) **11**, 13, **14**, 152, **153**, **219**
Endosperm 100, **101**, 225, **226**, 234
Endosymbiosis 15, **16**
End-product inhibition 157
Environmental persistence 236
Enzyme 12, 114, 149, 152, 154, **155**, 156, 178
Enzyme reaction 155, **157**
Epicotyl hook 204
Epidermal tissue 21
Epidermis **23**, **25**, **26**, **27**, **28**, **29**, 45
Equinox **198**
ER (endoplasmic reticulum) 13
Essential amino acids 150
Essential oils 236
Ethanol 184, **186**, 225
Ethephon 230
Ethylene 215, 221, **229**, **230**, **231**, **232**
Etiolation 204, **210**
Eukaryotic cells 15, **16**
Evaporation **133**, 138
Evaporative cooling **139**
Evening primrose **247**
Evergreen leaf 77
Exodermis **46**
Extensibility, cell wall 216
Extensin 6, 7

F₀ subcomplex 192
F₁ subcomplex 192
FADH₂ 180, 186, **187**, **188**, **190**, **193**
False hellebore **252**, 253
Far-red light 199, 200, 203, 204, **206**, 208
Fats 13, 144, **145**
Fatty acid 10, 144, **145**, **146**
Feedback inhibition 157
Feeding deterrent 241, 243
Fermentation 184, **186**, 225
Fertilization 100, **101**
Fertilizer **122**, 123
Fiber caps 17
Fiber cells 2, 3, 7, 9, **24**, **25**, 65, 79
Fibrous root system 36, **37**
Field capacity **109**, 128
Filament 91
Flacid guard cells 135, 233
Flannel bush 103, **104**
Flavin adenine dinucleotide–see FADH₂
Flavonoids 242, 244, 245
Flax 24

Floral stimulus protein–see Florigen
Florigen 202, **203**, **204**, **205**, 206
Flower 90, **91**
Flowering 195, **205**, **206**, 223
Food chain 158
Foxglove 241
Frankia 51
Freezing **214**
Fresh water 131
Frost **212**, **213**
Fructan 180, 182, **185**
Fructose **143**, 180
Fruit ripening **179**, 229
FT protein 202
Funiculus **102**
Fusiform initial 64

GA–see gibberellic acid
GA inhibitor **225**
Galactolipid 159
Galactose 10
Gallery, mountain pine beetle **239**
Gamete 91
Garner, Wightman 195
Gas exchange 71
Geiger counter 170
Gene 12, **153**
Generative cell 100, **101**
Germination 204, **208**, 225, **226**, 231, 234
Gibberella fujikuroi 221
Gibberellic acid (GA) 215, 221, 223, **224**, **226**, 234
Gladiolus 60
Glandular hair 236, **237**
Glucose **3**, **143**, 144, 156, 171, 180, **185**
Glyceraldehyde-3-phosphate (G3P) **172**
Glycerol 10, 144, **145**, 146
Glycolate **173**, 174
Glycolipid 10
Glycoprotein 13, **14**
Glycoside 240
Glycolysis 180, 182, 184, **185**, **193**, 194
Glyphosate 242
Golgi apparatus 6, 13, **14**, 152
Golgi vesicles 4
Goodyear, Charles 241
Grana 15, 159
Grand Rapids lettuce 201, 204, **208**
Grass flower **92**
Gravitational water **109**
Gravitropism **218**, **219**, **220**
Great Basin 251

Green Revolution 225
Ground tissue 21
Groundwater 131, 132
Growth **137**
Guanine **148, 149**, 226
Guard cells 15, 26, 116, 135, **136**, 233
Guttation 49, **50**
Gymnosperm 30, 89
Gypsum 118

H$^+$-ATPase **12, 126**, 216
H$_2$O–see water
Hardwood 65, 66, **67**
Harvest index 225
Hd3 protein 202
Heartwood 66, **67**, 247
Heme 154
Hemicellulose 4, 7, **14**
Herbicide 222
Heroine 251
Heterocyclic ring **249**
Hexose phosphate 182
Hong Kong 250
Hormones 215
Hummingbird 100
Humus 106, 107
Hydraulic lift 128, **129**
Hydrogen (H) **137, 143**
Hydrogen bond 3, 5, **138, 139**, 141, **142**, 149, **154**
Hydrogen ion (H$^+$) 166
Hydrolysis 143
Hydrophilic **10, 151**
Hydrophobic 10, 11, 150, **151**, 159, 161, **162**, 163
Hydrophyte 84, **85**, 231
Hydroxyl ion (OH$^-$) 109, **110, 143**, 242
Hygroscopic water **109**
Hypocotyl hook 204
Hypodermis 46, 78

IAA (indole-3-acetic acid)–see Auxin
Ice crystals **212, 213**
Immobile nutrient 119
Imperfect flower 92
Incomplete flower 92
Indole-3-acetic acid (IAA)–see Auxin
Induction 207, **210**
Infection thread **51**, 52
Inferior ovary **91**, 92
Inflorescence 94, **95, 96**
Infrared (IR) radiation 158

Inhibitor, enzyme 156, **157**
Initial 20, **41**, 64
Initiation 207, **210**
Inner bark 66, **67**, 68, 70
Inner matrix–see Matrix, mitochondrial
Inner membrane, mitochondrial 180, 186, 189, **190**, 194
Inorganic phosphate–see Phosphate (PO$_4^{3-}$)
Inquisition 250
Insecticide 241, 248
Integral protein 11
Intercalary meristem **19**, 21, 222
Intercellular space 11, **212, 213**, 231
Intermembrane space **189, 190, 191**
Internode **19**, 55, **56, 62**, 222, 223
Interveinal chlorosis **120**
Ionic bond 141, **142**
Ionizing radiation 158
Iron (Fe) 120, 154, 165, 189
Irrigation 131, **132**, 133
IR–see infrared
Isoprene 235, **236**

K$_2$O **122**, 123
Keel 98, **99**
Keratin 152
Krebs cycle 180, 184, 186, **187**, 186, 188, 190, 193, 194

Lactate or Lactic acid 184, **186**
Large-porous wood **66**
Larkspur **252**, 253
Lateral meristem 21
Lateral root **48**
Latex 241, **250**
Laticifer 241
Latitude 198, **199**
Leaf area 133
Leaf color **246**
Leaf elongation **19**
Leaf or Leaves 71
Leaf primordium 78
Leaf surface area **133**
Leaflet 71, **72**
Leghemoglobin 53
Legume 38, 52
Lemma 92
Lenticel 68, **70**
Light reactions of photosynthesis–see Photochemical reactions of photosynthesis

Lignin 7, 8, 47, 242, **243**
Lily 60
Liming 110
Lipid 9, 13, 144
Lipid bilayer 9, **10**, 11, 159, **162**, 211
Loam soil 108
Lodgepole pine 238
Lodging 225
Lodicules **92**
Long-day plant 196, **200**, 201, **205**
Lumen 13, 15, 152, **164**, 165, 166, **167**
Lupine **252**, 253
Lycopene **229**, 241

Macromolecules 141
Macronutrients 113, **122**
Magnesium (Mg) 117, 162
Maize 93, **223**, 233
Malate or Malic acid 13, **175**, 176, **177**, 233
Maltase 225
Maltose **226**
Mammalian toxicity 220, 236, 241
Manganese (Mn) 121, 165
Manure 108
Margulis, Lynn 15
Maryland Mammoth soybeans 196, **197**
Matrix, chloroplast (stroma) 15
Matrix, cytoplasmic 1
Matrix, extracellular 3
Matrix, mitochondrial 13, 180, 186, **190**, 191, 194
Maturation zone **42**, 43
Mechanical disturbance 230
Megapascal (MPa) 128
Membrane lipids 10
Membrane proteins **10**, 11
Membranes 9, 13
Meristem 18, 40, **80**
Mesophyll **17**, 21, 29, 80, 82, 88, **174**, 175, 180, **181**, 213
Messenger RNA–see mRNA
Metabolism 12
Metaxylem **49**
Methane 229
Methionine 118, 229
Methyl group 146
Mg^{2+}-ATPase **156**
Microfibrils 3, 4, 5
Micronutrients 119
Micropyle 100
Midden 103

Middle lamella 4, 6, 7, **8**, **9**, **25**, 117
Milkweed **240**
Mineral particles 106, **107**
Mitochondria 1, 13, **14**, 15, **16**, **25**, 186, 189, **193**, 194, 209
Mobile nutrient 114
Molecule **142**
Monarch butterfly **240**
Monecious species 92, **93**
Monkshood 253
Monocot 7, 36, 78, **80**, 83
Monomer 143
Monoterpene **236**
Morphine **249**, 250
Morphology 1
Mountain pine bark beetle 238, **239**
mRNA (messenger RNA) 151, 152, **153**
Mucigel 41
Multiple fruit 92
Mycorrhizae 118

NAD^+ 183
NADH 180, 183, **184**, **185**, 186, 187, **188**, 190, 193
$NADP^+$ (nicotinamide adenine dinucleotide phosphate) 165
NADP reductase **164**
NADPH 137, 158, **159**, 164, 165, 166, 172, 177, 188
Natrophilic 121
Nectar 98
Nectar guide 98, 245
Nectary 94, **95**
Needle 78, **80**, 237
Neem tree 241
Net photosynthesis **179**
Neurotoxin 248
Neurotransmitters 248
Nickle (Ni) 121
Nicotinamide adenine dinucleotide phosphate–see NADP
Nicotine 248, **249**
Nitrate (NO_3^-) 13, 111, 116
Nitrogen (N) 78, 114, **115**, **116**, 122
Nitrogen (N_2) 49, 53, 122, **232**
Nitrogen fixation **51**, 53, 54, 121
Nitrogenase **51**, 53
Nodal Roots **37**, 38
Node 55, **56**, **62**, 72
Nodule **52**, 53
Nodule primordium **51**, 52
Noncompetitive inhibitor 157

NPK ratio 122
Nuclear envelope 13
Nucleic acid 147
Nucleotide 147, **148**, **149**
Nucleus 1, 12, 25, **152**, **153**
Nutrient uptake **112**
Nutrients 10, 111, 114

OAA–see Oxaloacetic acid
Ogalla aquifer 131, **132**, 133
Opiate 251
Opium 249
Opium Wars 250
Organelles 12
Organic acid 13
Organic chemistry 141
Organic matter 106, 107, **108**
Organic molecules 141
Organic soil 107
Organogenesis 227
Osmosis 88, 124, **125**, 126, 127, 129, 233
Osmotic potential 129
Osmoticum 121
Outer bark **67**, 68
Ovary 89, **91**, 92, 100, 102
Ovule 91, 100, **101**, **102**
Oxaloacetic acid (Oxaloacetate) (OAA) 175, 186, 188
Oxidation-reduction reaction 121, 155, 188, **189**, 190, 191
Oxidative phosphorylation 190
Oxygen (O) **137**
Oxygen (O$_2$) 10, 71, **164**, **165**, 166, 171, 173, 174, 189, 194, 231, **232**
Oxygenase 171, **173**

P$_2$O$_5$ **122**, 123
P680 **164**, 165
P700 **164**, 165
Palea **92**
Palisade parenchyma 81
Palmate leaf 72, **73**
Pando 97
Panicle **95**, 96
Papavarine 251
Paper chromatography 169
Paragoric 251
Parallel veins 82, **83**
Parenchyma cells 21, **33**
Passive transport 124, 129
Pectin 4, 6, 7, **14**, 117
Pectinase **229**

Pedicel 94, 95
Peduncle 94
Penstemon **99**
People's Republic of China 250
PEP–see Phosphoenol pyruvate
PEP carboxylase **175**
Peptide bond 149, **150**, 153
Perennials 36
Perfect flower 92
Perforation plates 30
Perianth 90
Pericarp **226**
Pericycle 47, **48**
Peripheral protein 11
Permanent root 37
Permanent wilting point **109**, 135
Petal 89, **91**
Petiole 23, 72, **73**, 75
3-PGA–see 3-Phosphoglycerate
pH (acidity or alkalinity) 110
Pharmocology 241
Phenol 242
Phenolic monomer 244
Phenolics 235, **242**, **243**
Phenyl ring **242**
Phenylalanine 242
Pheromones 238
Phloem **17**, 21, 30, **33**, 48, 68, 88, **174**
Phloem fibers **68**
Phloem ray **68**
Phloem tissue 30
Phosphate (PO$_4^{3-}$) 10, 112, 118, 144, 147, **148**, **149**, 156, 165, 166, **167**, 171, 175, 180, **181**, **182**, **183**, 185, 190, 191
Phosphate bond 12, **172**, 176
Phosphoenol pyruvate (PEP) **175**, 176
3-Phosphoglycerate (3-Phosphoglyceric acid) (3-PGA) **167**, **170**, 171, **172**, **173**, 174, 180
Phospholipids **10**, 118, 144
Phosphorus (P) **115**, 118, **119**, 122
Phosphorylation 156
Photochemical reactions of photosynthesis 158, **159**, 163, 166, 168, **177**
Photoperiod **199**
Photoperiodism 196
Photophosphorylation 165
Photorespiration **173**, 174, 175, 176, 180
Photosynthate 85, 87
Photosynthesis 15, 71, 81, 121, 124, **130**, 133, 135, 136, **137**, 156, 158, 168, **173**, **175**, 179, **181**

Photosystem I **164, 165, 167**
Photosystem II **164, 165, 167**
Phototropism 216, **217**
Phragmoplast 6
Phytochrome 199, 200, 204, **208**, 231
Phytochrome far-red **201**, 202, **205, 206**
Phytochrome red **201**, 202, **205, 206**
Phytomer 37
Pigment 200
Pine needle 237
Pinnate leaf 72, **73**
Piperidine 253
Pistil 91
Pistillate flower 89, **91**
Pit membrane 30, **32**
Pitcher plant **78**
Pith 21, 47, **49**
Plant cell 1, **2**
Plasma membrane 1, **8**, **10**, 209
Plasmalemma 8
Plasmid DNA 15
Plasmodesmata 2, **11**, 17, **25**
Plastids 15, **25**, 232
Pod **102**
Poinsettia 94
Polar molecule 137
Polar movement 216, 219
Polar nuclei 100, **101**
Polarity **137**, **138**
Pollen 91, **97, 98**, 221
Pollen grain 100, **101**
Pollen guide **99**, 245
Pollen tube 100, **101**
Pollination 62, 97, **98**, 100, **101**, 104, 244
Pollinator 98
Polygalacturonase **229**
Polymer 143, **144**
Polyterpene **236**, 241
Poppy **250**
Posttranslational modification 13
Potassium (K) 26, **115**, 116, **117**, 122, 233
Precipitation 131
Predawn water potential 134
Primary cell wall 2, 6, 7, 9, **25**
Primary growth 18, 62, **63**
Primary protein structure 152, **154**
Primary root 37
Product, enzymatic 155, **157**
Prokaryotic cells 15
Protein 9, 13, 149
Protein channel **126**
Protein complex 152, **154**

Protein structure **154**
Protein synthesis 13, 152, **153**
Proton (H^+) 12, 15, 109, **110**, **112**, 164, 165, 167, 189, 190, 191
Protoplast 1, 6, **25**, **212**
Pubescence 73, **76**
Pyrethrins 236
Pyruvate **175**, 180, 183, 184, **185**, 186, 187, 188

Quaking aspen
Quaternary protein structure 152, **154**
Quiescent center 20, **41**
Quinolizidine 253

Raceme 94, **96**
Rachis 94
Radicle 37
Ray 65
Ray flower 95, **96**
Ray parenchyma 64
Reaction center **163**, 165
Receptacle 89
Receptor molecule **163**
Recharge 132
Red light 199, 200, 203, 204, **206**, **208**
Red-to-far-red light ratio 204, **209**
Reducing power 188, **190**
Relative humidity 131
Resin 236, 238
Resin duct 236, **237**
Respiration 13, 88, 121, 178, **179**, 184, 193
Respiration rate **179**, 229
Reticulate veins 82, **83**
Rhizobium bacteria 51
Rhizome 22, 57, **58**
Ribonucleic acid–see RNA
Ribose 147, 148, **183**
Ribosomal RNA 151, 152
Ribosome 13, 152, **153**
Ribulose-1,5-bisphosphate (RuBP) 171, 172, **173**
Ribulose-1,5-bisphosphate carboxylase-oxygenase–see Rubisco
Ribulose-5-phosphate (Ru5P) 172
RNA (Ribonucleic acid) 12, 147, **148**
Root 36, **45**
Root cap 19, 40, **41**, **42**, 218, **220**
Root hairs 29, **43**, **44**
Root nodule 52, **53**
Root pressure 49

Root tip **43**
Rosette 55, **56**, 72, 223
Rotational catalysis 191, **192**
Rough ER (endoplasmic reticulum) 13
Rubber 241
Rubisco 152, 155, 156, 171, **172**, **173**

Sainfoin **247**
Salinization **131**
Samara **102**
Sand 106, **107**
Sapodilla tree **240**
Sapwood 66, **67**, 238
Saturated fat **144**
Saturated fatty acid 144, **146**, 147, 211
Saturated soil **109**
Scale leaf 78
Sclerenchyma cells 23
Sclerid 2, 24, **25**
Sclerophyllous leaf 77
Secondary cell wall 2, 6, 7, **9**, 25
Secondary cortex 68
Secondary growth 18, 62, **63**
Secondary metabolism 235, 248
Secondary phloem 33, 64, 67
Secondary protein structure 152, **154**
Secondary xylem (wood) 21, 64
Secretory vesicles 13
Seed 91, **101**
Seed dispersal **102**, 103, 244
Seminal root 37
Senescence 178, **228**, 229, 244, 246
Sepal 89
Sesquiterpene 236
Sessile leaf 72, **74**
Shade **207**
Shade leaf **83**
Sheath, leaf **19**, 72, 74
Shikimic acid pathway 242
Short-day plant 196, 200, 202, **203**, 205
Side chain 150, **151**, 152, **154**
Sieve area 31, 32
Sieve cells 31
Sieve element 31
Sieve tube 12, 31, **33**
Silicone (Si) 119
Silt 106, **107**
Simple leaf 71
Sink, photosynthate 87, 88
Smooth ER (endoplasmic reticulum) 13
Sodforming grass **58**
Sodium (Na) 121

Softwood 65
Soil nutrients **115**
Soil structure **108**
Solar spectrum **206**
Solute 124, **125**, 127, **128**
Solute potential 127
Solvent 124, **125**
Source, photosynthate 85, 87, 88
Sperm cell 100, **101**
Spike 94, **96**
Spine 55, **57**, 77
Spongy mesophyll or parenchyma 81
Spreader 138
Springwood 65
Spur 94, **95**
Stamen 89, **91**
Staminate column 98, **99**
Staminate flower 89, **91**
Standard petal **99**
Starch 3, **4**, **22**, 88, 144, 180, **181**, **182**, 185, 225, **226**, 229
Stele 47, **49**, 218
Stem 55, 56, **61**, 63
Stem elongation 121, 204, **209**, 222
Stigma 91, 100, **101**
Stipules **73**, 75
Stolon 57, **58**
Stomata 26, 27, **28**, 71, 78, 84, 124, 130, 133, **134**, 135, 136, 137, 171, 174, 176, 177
Stomatal closure **136**, 173, 233
Stomatal complex 26
Stomatal conductance **134**
Storage parenchyma 46
Stratification 207, **211**
Streptavidin **192**
Stroma 15, **159**, 164, 166, **167**, 171, 244
Style **91**, **101**
Suberin 46, 47, 112
Suberization 43, **230**
Subsidiary cells 26
Substrate, enzyme 154, **155**
Subunit, protein 152, **154**
Succulent 74, 76, 84, 176
Sucrose 85, 87, 88, 143, 171, 180, **182**
Sugars ($C_6H_{12}O_6$) 158
Sulfate (SO_4^{2-}) 10
Sulfolipid 10, 159
Sulfur 118, 189
Summer solstice **198**
Summerwood 65
Sun fleck **207**

Sun leaf 82, **83**
Superior ovary **90**, 92
Surface tension **138**
Surfactant 138
Symbiotic relationship 15, 49
Symplast 11, 47, 113, 129
Symplastic movement 87, **113**

Tannin 242, 245, **247**
Taproot 36, **37**, **38**, **39**, 87
TCA cycle–see Tricarboxylic acid cycle
Temperature **176**, 180
Tendril **59**, 77, 79
Tensile strength 138, **139**
Tepal 90
Terpene glycosides **240**
Terpenes 235, **236**
Tertiary protein structure 152, **154**
Tetrapyrrole ring 161
Tetraterpene 241
Thermal stability 138
Thigmomorphogenesis 230, **231**
Thigmotropism 218
Thylakoid membranes 15, **159**, 161, **162**, 163, **164**, 165, 244
Thymine 147, **148**, 149, 226
Tiller, grass 57, **58**
Tonoplast 12, 209
Torus 30, **32**
Tracheid 30, **31**, 65, **66**
Trans double bond **146**
Trans fatty acids 147
Transcription 12, 121, 152, **153**
Transfer RNA–see tRNA
Translation 12, 152, **153**
Translocation 85, 87, **88**, 171
Transpiration 73, 84, **130**, 131, **133**, **134**, 135, 138, 140, 176, 177
Transport vesicles 13
Trapping leaf 77, **78**
Tricarboxylic acid (TCA) cycle 186, **187**
Trichome 24, 27, **29**, 77, 84
Trifoliolate leaf 40
Triglyceride 144
Triose phosphate 180, **181**, 182
Tripped flower **99**
Triterpene 240, 241
tRNA (transfer RNA) 151, **153**
Tropism 216
Tube cell 100, **101**
Tuber 60
Tuberous root 59, **60**

Turgid, Turgidity 23, 127, 135
Turgor **125**, **136**
Turgor pressure **127**
Twining shoot **59**
Type II diabetes 147

Ultraviolet (UV) absorbance **104**, **247**
Ultraviolet (UV) radiation 158, 245
Umbel 95
Unifoliolate leaf **40**
Unsaturated fat 144
Unsaturated fatty acid 10, 144, **146**, 147, 211
Uracil 147, **148**
Urea 121
Uronic acid 4
UV–see ultraviolet

Vacuole **12**, 127, 176, **177**, 180, 244, **245**, **246**
Vascular bundle **17**, **22**, 60, **61**
Vascular cambium 21, 28, 62, 64, 66
Vascular tissue 21, 28
Vegetative clone 97
Vegetative propagation or reproduction 57, 96, 97
Veins 81
Venation, leaf **83**
Venus Flytrap **78**
Vernalization 205, 206, 207, 209, **210**, 224
Vessel element 30, **31**
Viceroy butterfly 240
Visible light 158, **160**, 207
Vivipary **233**, 234
Vulcanization 241

Water 124, 137, **138**, 158, **165**, 189, 194
Water consumption 131
Water movement **126**, **136**
Water potential 127, **128**, 129, **130**, 134, 135, 136
Water potential gradient **128**
Water storage tissue **84**
Water uptake **43**, 125, 130, 135
Water-holding capacity 106, 107
Waterlogging 231, **232**
Wilting 127
Wind pollination 97
Winter solstice **198**
Winter-hardiness 117, 209, 212, **213**, 214
Witches' broom **227**

Withering, William 241
Wood 8, **65**, **66**, **67**
Woody root **50**

Xerophyte **84**, 133
Xylem 2, 7, 8, 11, **17**, 30, **31**, 48, **174**

Xylem element **49**, 129
Xylem parenchyma **113**, 129

Zeatin 226
Zinc (Zn) 121
Zygote 100, **101**